Toxicology

for

Non-Toxicologists

Mark E. Stelljes, Ph.D.

Government Institutes
Rockville, Maryland

Government Institutes, a Division of ABS Group Inc.
4 Research Place, Rockville, Maryland 20850, USA
Phone: (301) 921-2300
Fax: (301) 921-0373
Email: giinfo@govinst.com
Internet: http://www.govinst.com

03 02 01 00   5   4   3   2

Library of Congress Cataloging-in-Publication Data

Stelljes, Mark E., 1957-
    Toxicology for non-toxicologists / Mark E. Stelljes.
       p. cm.
    Includes bibliographical references and index.
    ISBN 0-86587-611-8 (pbk.)
    1. Toxicology–Popular works. I. Title.

RA1213.S73 1999
615.9–dc21

                                          99-051974

Contents

Chapter 1
Introduction 1

Chapter 2
Everyday Applications of Toxicology 5

Chapter 3
Essential Concepts of Toxicology 23

Chapter 4
Types of Toxic Effects 39

Chapter 5
Estimation of Toxic Effects 55

Chapter 9
Communicating Risks to the Public 121

Chapter 10
New Approaches in Toxicology and Risk
Assessment 133

Appendix A
Additional Reading 147

Appendix B
Glossary 151

List of Figures and Tables

Preface

To some of us, the term "toxicology" is typically heard in association with autopsy stories on the news—"police are still awaiting results from the toxicology tests before they will say the death was drug-related." But what are toxicology tests, and how are results obtained and interpreted? On a larger scale, we can think about pesticide testing on fruits and vegetables, and wonder about the sort of tests that are done on produce. Or as we decide whether to install a water filter, we can wonder if the water quality is safe for drinking, and how a "safe" amount is determined. All of these questions and issues fall under the discipline of toxicology.

The purpose of this book is to present basic concepts in toxicology in a way that is applicable to everyday life. Toxicology is a multidisciplinary field that incorporates aspects of biology, chemistry, physics, medicine, engineering, genetics, risk assessment, and other fields. Books on the subject are typically geared toward professionals in these fields. This makes undertaking a study of toxicology daunting to those without this specific background. However, this complexity also indicates toxicology's broad scope of impact. Because toxicology affects us, it is important to understand some basic concepts of the discipline.

Risk assessment and risk communication are other areas related to toxicology. Few textbooks and manuals are available on these topics, and they are rarely taught in university curriculums. Yet we often hear about some chemical or other being risky or increasing the risk of cancer, but we are without the ability to evaluate the risk.

This book is intended to bridge the gap between toxicological research and applications that impact our lives. Some technical information is presented because it is essential for understanding some of the concepts, but in general the jargon is kept to a minimum. As such, this book fills a vacancy in the toxicology literature.

Although this book is primarily intended for professionals in the environmental field, such as managers, engineers, and geologists, it is also designed for a more general audience. It is hoped that this book will enable you to answer many of your own questions, such as:

- Which of the two conflicting news stories about whether a chemical causes cancer should I believe?
- How do I know if my water supply is safe?
- Will I get cancer from chemicals, and how can I avoid it?
- Why are animals still used in toxicology testing?
- Should I be concerned about emissions from that factory in my neighborhood, or that truck exhaust ahead of me?

This book is organized into ten chapters. Chapters 1 through 4 present an overview of how toxicology impacts our everyday lives, the essential concepts of toxicology, and the types of toxic effects that chemicals can have. Chapters 5 through 7 present a more detailed discussion of core concepts in toxicology and the basics of animal toxicological testing and extrapolation to humans. Chapters 8 and 9 discuss risk assessment and communication. Chapter 10 introduces some relatively new issues in the discipline that will likely grow in technical importance and overall level of impact on our lives in the future. The select list of additional reading (Appendix A) provides a launching point to more information on some of the topics discussed in the text. A glossary of terms is provided in Appendix B. Because concepts in later chapters rely on terms and information presented earlier, chapters are best read in the order presented.

About the Author

Dr. Mark Stelljes received a Bachelor's degree in Zoology from the Ohio State University in 1978, a Master's degree in Wildlife Biology from the University of Missouri-Columbia in 1982, and a Ph.D. in Environmental Toxicology and Pharmacology from the University of California–Davis in 1990. Between degrees, he worked in the Peace Corps in Tunisia conducting animal population surveys at a new national park. He has spent the last decade in the field of environmental consulting, where he has been primarily involved in conducting risk assessments to evaluate the need for remediation (e.g., cleanup) at hazardous waste sites, with the goal of ensuring protection of humans and the environment. He has authored many papers in the field of environmental toxicology and risk assessment, and has lectured in these fields to undergraduates at both UC Davis and UCLA. He currently serves as the director of risk assessment and toxicology services for SECOR International Incorporated, in Concord, California. SECOR is an environmental consulting firm headquartered in Washington with over 40 offices in the United States and several more in Canada and the United Kingdom. He is also active in the Society of Environmental Toxicology and Chemistry, a nonprofit organization dedicated to improving research, education, and outreach in the fields of environmental toxicology and chemistry. He served on the Board of Directors for the Northern California chapter of this society from 1997 to 1999, and served as Secretary, Vice-president, and President during this period. He currently lives in the San Francisco Bay Area with his wife and daughter.

Acknowledgments

Many people over the years have helped shape my views and knowledge of toxicology and risk assessment. Without them this book could not have been written. I wish to thank Les Williams and Gayle Edmisten Watkin for assisting my professional growth and understanding of the field. I would also like to thank Dr. Robert Breitenbach for urging me to consider a career in environmental toxicology back when I was in graduate school. I also would like to thank Dr. David Jeffrey, Daniel Lee, and Ivy Inouye for their contributions to this book through their association with me at SECOR International Incorporated. I also must thank Jim Vais and others at SECOR for allowing me to prepare this manuscript. Finally, I would like to thank my wife, Kathryn, for her valuable editorial review of the manuscript and her patience and understanding for all the weekends and evenings I was occupied with it. Without her support, this book would not have come to fruition.

I

Introduction

Toxicology is the study of poisons, or the study of the harmful effects of chemicals. Some of you may be familiar with toxicology from watching TV shows such as *Quincy*, which dealt with forensic toxicology. Others may know only what you have read in newspapers or have seen on the news. But television shows and media reports typically focus on controversial subjects and may not provide enough factual information for the public to make informed decisions. This book will help you sift through all this information and explain the value of understanding the basic concepts in toxicology that are relevant to everyday life.

Although this book is intended for non-scientific audiences, some of the information may seem somewhat esoteric. However, whenever possible, common examples are provided to allow the reader to understand the context of the information and its relevance to everyday life.

Why should you care about toxicology? Isn't it enough to know that you don't want pesticides on your food or chemicals in your water?

Chemicals are everywhere in our environment. All matter on our planet consists of chemicals. We are made up of a few thousand different types of chemicals, some of which are considered toxic. The vast majority of these chemicals are natural; in fact, the most potent chemicals on the planet are those occurring naturally in plants and animals. Natural chemicals are sometimes presented by the media as being "safe" relative to manufactured chemicals. As a result, they may also be considered "safe" by much of the public.

For example, "organic" produce and livestock are becoming more popular. But does this mean pesticides are unsafe? Actually, manufactured pesticides used on crops and animals are heavily regulated and rarely contain enough chemicals to be harmful at typically encountered levels. This book is designed to enable you to make your own conclusions about health hazards from chemicals in the environment, and to be a more informed professional and consumer.

1

Also, lack of knowledge about a topic often leads to unnecessary fear of the unknown. On the other hand, ignorance can lead to disregard of the potentially catastrophic consequences of our actions. The following discussion of historical events gives an example of the latter concept.

For decades the Roman Empire was the crown jewel of the world. The expanse of land controlled by the Romans grew at an amazing rate during this period, and their rule led to the development of irrigation systems, roads, public sporting events, and a government that, at least in part, helped improve the lot of its citizens. With the success of the empire and the expansion of other societies came the trappings of wealth. However, this wealth was retained in the ruling class. Great feasts were held daily in which the ruling class, especially the emperors, ate tremendous meals and imbibed wines from all over the world. Some of the poorer quality wines were augmented with flavor enhancers before they were consumed by the aristocracy. This wine was served in the best goblets, made by master craftsmen and collected from around Greece and other European countries. The meals were served in the best bronze and copper pots. Emperors such as Nero, Claudius, and Caligula were perhaps among those who took this feasting to its greatest extent.

Meanwhile, the common folk were unable to afford the great wine and lovely goblets, and instead were only able to buy cheap wine flasks and poor quality wine. They were also unable to afford the bronze and copper pots for cooking, and were only able to buy cheap metallic pots.

After several years, the rulers began to behave strangely. Claudius started forgetting things and slurring his speech. He also began to slobber and walk with a staggering gait. Many of the decisions he made adversely impacted the empire, and there was little basis for these decisions. Eventually, he was replaced as emperor because he could no longer function as a ruler. In another famous story, the mad emperor Nero fiddled while his city burned. He became insane while ruling, and his actions had an adverse impact on the empire. Caligula became sexually depraved, and suffered a mental breakdown. The ruling class seemed plagued by neurological diseases. Eventually, the repeated succession of apparently incompetent rulers began the decline of the Roman empire.

For hundreds of years the actions of these rulers were not understood. It was a mystery why similar diseases did not appear with nearly the same frequency in the common class. Why was it that the aristocracy had such a high rate of neurological disorders?

Historians interested in Roman architecture and society soon discovered some facts about Roman and Greek manufacturing that at first seemed unrelated to these events. The Romans and Greeks would typically coat bronze and copper cooking pots and goblets with lead to prevent copper or other metals from being dissolved into food or drink. The flavor of poor wines was enhanced by adding lead compounds prior to being served to the aristocracy.

Writers of the Roman empire era noted that the excessive use of lead-treated wines led to "paralytic hands" and "harm to the nerves." Numbness, paralysis, seizures, insomnia, stomach distress, and constipation are other symptoms of lead toxicity. Now we know that lead is a neurological poison when even relatively low amounts are consumed. A theory was advanced that the neurological problems of the emperors may have been caused, at least in part, by lead toxicity. To test the theory, researchers prepared a liter of wine according to an ancient Roman recipe and extracted 237 milligrams of lead. Based on the known excessive habits of these emperors, it was estimated using risk assessment techniques that the average Roman aristocrat consumed 250 micrograms of lead daily. Some aristocrats, notably the emperors discussed above, may have consumed a gram or more of lead daily. Today, Americans that live in cities are known to have the highest lead exposures from a variety of sources, including leaded water pipes, lead-based paint, and food. Even with the concern about lead exposure by these people, the average city-dwelling American consumes only about 30 to 50 micrograms of lead daily. By this comparison, the Roman aristocracy consumed on average 8 times what a highly exposed city-dwelling American now consumes. Some emperors may have consumed 20 times more than city-dwelling Americans.

Despite the more obvious neurological effects, the more important effects of lead may have contributed to the decline of the Roman empire through contributing to a declining birth rate and shorter life span among the ruling class.

This example illustrates that general lack of knowledge about toxicology led to everyday practices that may have contributed to shaping our history. Similarly, toxicology has an impact on our lives today.

The foods we eat and the additives put into them are routinely tested for chemicals. Levels of pesticides are randomly measured in food. Our drinking water is purified to eliminate harmful chemicals. Air pollution is monitored, and many efforts are under way to improve our global air quality. All of these routine actions in today's world originated when the effects of various substances on our health and the health of wildlife and the environment became evident.

The study of the effects of chemicals on our health and the health of wildlife combines toxicology with risk assessment. Risk assessment evaluates the relative safety of chemicals from an exposure approach, considering that both contact with a chemical and the inherent toxicity of the chemical are needed to have an effect. We can't eliminate the toxicity of a chemical, but we can limit our exposure to it.

This is the general approach behind the development of most pesticides. Many chemicals are designed to target only insects, leaving the crops essentially free of pesticides. Even though the pesticides are toxic, there is very little risk from eating the crops because almost no chemical remains in the food.

The discussion of lead and risk assessment focuses on how a basic understanding of toxicology has applications in everyday life. The purpose of this book is to provide some basic concepts of toxicology, presented in such a way as to make them applicable and meaningful to non-scientists. It will probably be necessary for the reader to learn a few new terms and wade through a few simple graphs and illustrations, but in general the attempt is to provide some relevance to the information.

The book focuses on toxicology as it relates to humans. It presents some concepts of ecological toxicology, and risk assessment is further discussed. The book also illustrates how some information on toxicology is applied to real-world decisions.

Before getting into a discussion of toxicology, Chapter 2 presents several examples of how toxicology and risk assessment impact us in everyday life. The intent of the chapter is to provide a backdrop to the rest of the book, and show that several things we take for granted are based in part on toxicology.

Chapters 3 through 6 present the important concepts in toxicology. Chapter 7 lays the groundwork for a discussion of risk assessment by discussing how toxicological information compiled for laboratory animals is extrapolated to humans. Chapters 8 through 10 provide a discussion of the general approach to risk assessment, how it can apply in important ways, how it gets communicated to the general public, and new approaches that may be more accurate than the methods currently used to assess the toxicity of chemicals.

Most chapters contain relevant bibliographies. There is also a full bibliography at the end of the book.

Additional Reading

Gilfillan, S. C. 1965. Lead Poisoning and the Fall of Rome. *Journal of Occupational Medicine.* 7:53-60.

Nriago, J. O. 1983. *Lead and Lead Poisoning in Antiquity.* John Wiley and Sons, New York. 437 pages.

Waldron, H. A. 1973. Lead Poisoning in the Ancient World. *Medical History.* 17:391-399.

2

Everyday Applications of Toxicology

Is the food I eat safe? What about the air I breathe and the water I drink? Is it as bad as the media says? Should I worry, and if so, when? These questions arise often in this age of high chemical use. We use more chemicals on our crops than ever before. As more countries become industrialized, air pollution becomes a pressing issue. Are we filling in all of our wetlands? How will this impact our water supply?

This chapter hopes to demonstrate that, although we are often guinea pigs with new chemicals, it is perhaps surprising that we are only rarely affected by chemicals released into the environment. But when we are, it is big news, so these rare events tend to occupy our minds. This is like reading about a plane crash—many are affected at one time, but the method of travel is statistically safe relative to other forms of transportation.

In the 16th century, Paracelsus said, "All things are poison and nothing [is] without poison. Solely the dose determines that a thing is not a poison." This is still true today. A dose is defined as the amount of a chemical entering the body, and is a function of the amount of and exposure to that chemical. Sometimes we are exposed to a small amount and get extremely ill (e.g., food poisoning), while at other times we are exposed to a lot and don't notice a thing (salt).

Paracelsus' quotation is key to understanding what toxicity means. Knowing that a chemical is toxic does not automatically imply that adverse effects will result from exposure to any amount. Most of us can tolerate reasonable amounts of table salt without fear of adverse effects. However, ingesting too much salt can cause heart toxicity, or disrupt the water balance in our cells, blood, and kidneys. Toxicity can be defined as the degree to which a chemical inherently causes adverse effects.

As these above examples imply, and as we will see in Chapter 3, toxicity is a function both of the inherent actions of a chemical and the amount of a chemical to which we are exposed. One good example is prescription drugs. A doctor prescribes sleeping pills for someone with insomnia or other sleeping problems. Taking the prescribed amount lets the patient sleep yet does not cause adverse effects. However, if the whole bottle of pills is taken, toxic effects, including death, can result (e.g., suicides). The chemical ingredients in sleeping pills are toxic, but only in doses above a certain amount. How then do you categorize sleeping pills? Are they toxic, or not? This example illustrates that the concept of toxicity is not clear-cut.

For an understanding of the relative danger associated with a chemical, inherent toxicity must be linked to some level of exposure. We have developed a relative toxicity scale for chemicals, depending on the amount of a chemical necessary to cause adverse effects. This scale is shown in Table 2.1, along with examples for each category. This scale can be used to assign relative toxicity to chemicals. For example, cyanide is more toxic than table salt because toxic effects can result from much lower amounts of cyanide. This implies that cyanide is more dangerous than table salt. However, we are not usually exposed to cyanide, but we are exposed to table salt daily. Which chemical is more dangerous for the average consumer? This is a difficult question to answer, and we will address the question in detail in Chapter 8, Risk Assessment.

Table 2.1. Scale of Relative Toxicity

Category[a]	Concentration[b]	Amount for Average Adult[c]	Example Chemicals
Extremely Toxic	less than 1 mg/kg	taste	botulinum, tetrodotoxin, dioxin
Highly Toxic	1-50 mg/kg	7 drops – teaspoon	nicotine, strychnine, hydrogen cyanide
Moderately Toxic	50-500 mg/kg	teaspoon – ounce	DDT, acetaminophen, aspirin
Slightly Toxic	500-5,000 mg/kg	ounce – pint	salt (NaCl), malathion
Practically Nontoxic	5,000-15,000 mg/kg	pint – quart	ethanol, MTBE, daminozide (Alar)
Relatively Harmless	more than 15,000 mg/kg	more than 1 quart	water

mg/kg = milligrams of chemical per kilogram of body weight.

[a] Based on ability to cause death from a single dose (i.e., acute exposure).
[b] Amount needed to cause death from a single dose.
[c] Assumed to weigh 70 kg (155 lbs).

The issues discussed above illustrate how toxicology impacts our everyday lives. The rest of this chapter expands on these issues and separates them based on how, and to what extent, exposure can occur. This chapter first discusses food toxicology, including manufactured chemicals (additives) and natural ingredients. Then examples of air pollution are presented, followed by episodes of water pollution. The impact the media has on our impression of these events is then discussed. This chapter ends with an important discussion about our lack of knowledge of toxicology. Are we perhaps like the Romans, sitting on a time bomb that we can't see, or are we generally safe from our environment?

Food Toxicology

Chemicals are everywhere in our environment. All matter on our planet consists of chemicals. The human body is made up a few thousand different types of chemicals, some of which are considered toxic. For example, we have hydrochloric acid in our stomachs that assists in food digestion. Our stomach lining is designed to protect us against the amounts of this acid secreted into the stomach. This lining is affected when we get an ulcer. Yet we typically don't notice this acid, and suffer no toxic effects from its presence. Animals other than humans have extremely potent chemicals in their bodies, especially marine organisms (e.g., puffer fish). A list of very potent chemicals found in marine organisms is shown in Table 2.2.

Table 2.2. Toxic Chemicals Produced by Marine Organisms

Chemical Name	Concentration[a]	Source of Chemical
Tetrodotoxin	0.008 mg/kg	Puffer fish[b]
Saxitoxin	0.009 mg/kg	Shellfish[c]
Protein toxin	0.0116 mg/kg	Sea urchin (*Lytechinus variegatus*)
Equinatoxin	0.033 mg/kg	Sea anemone (*Actinia equina*)
Nematocyst toxin	0.04 mg/kg	Fire coral (*Millepora alcicornis*)
Physalitoxin	0.14 mg/kg	Portuguese man-of-war jellyfish (*Physalia physalis*)
Protein venom	2.0 mg/kg	Rockfish (Scorpaena)
Murexine	8.1 mg/kg	Abalone (*Murex* sp.)

mg/kg = milligrams of chemical per kilogram of body weight.

[a] Amount needed to cause death from a single dose.
[b] Also produced by species of newts and frogs (amphibians) and octopi (mollusks).
[c] Produced by microorganism and subsequently ingested by shellfish; source of some "red tides."

Terrestrial species, including reptiles (e.g., rattlesnakes, cobras) and amphibians (e.g., frogs), also contain extremely potent chemicals. Examples of extremely potent chemicals present in reptiles and amphibians are shown in Table 2.3.

Plants are composed of different sets of chemicals than animals, and they contain some of the most potent chemicals on earth (e.g., digitonin from foxglove [digitalis] plants). Examples of very potent chemicals found in plants are shown in Table 2.4. Some of these same chemicals are also used for healing properties (digitalis is used as a drug for heart disease).

Table 2.3. Toxic Chemicals Produced by Reptiles and Amphibians

Chemical Name	Concentration[a]	Source of Chemical	Type of Organism
Cobra neurotoxin	0.0003 mg/kg	Cobra	Reptile
Crotalus toxin	0.0002 mg/kg	Rattlesnake	Reptile
Kokol venom	0.0027 mg/kg	Frog	Amphibian
Batrachotoxin	0.0028 mg/kg	Frog	Amphibian
Tarichatoxin	0.008 mg/kg	Newt	Amphibian
Bufotoxin	0.39 mg/kg	Toad	Amphibian

mg/kg = milligrams of chemical per kilogram of body weight.

[a] Amount needed to cause death from a single dose.

Table 2.4. Toxic Chemicals Produced by Plants

Chemical Name	Concentration[a]	Source of Chemical
Ricin	0.00002 mg/kg	Castor oil plant (*Ricinus communis*)
Digitoxin	0.07 mg/kg	Foxglove (*Digitalis purpurea*)
Amanitin	0.1 mg/kg	Deathcap mushroom (*Amanita phalloides*)
Curare	0.5 mg/kg	South American "arrow poison" (*Chondodendron tomentosum*)
Atropine	0.7 mg/kg	Jimson weed (*Datura* sp.), Deadly nightshade (*Atropa belladonna*)
Strychnine	1.0 mg/kg	Strychnos nux-vomica
Nicotine	1.0 mg/kg	Tobacco (*Nicotiana tabacum*)
Hydrogen Cyanide	2.0 mg/kg	Almond, apricot trees
Coniine	10 mg/kg	Poison hemlock (*Conium maculatum*)
Oxalic acid	5,000 mg/kg	Philodendron (*Philodendron* sp.)

mg/kg = milligrams of chemical per kilogram of body weight.

[a] Amount needed to cause death from a single dose.

Insects (e.g., brown recluse and black widow spiders) have yet different sets of chemicals, some of which are also very potent. Microorganisms also contain some very potent chemicals (e.g., botulinum toxin that causes botulism). Table 2.5 lists some examples of very potent chemicals found in insects and microorganisms.

Table 2.5. Toxic Chemicals Produced by Insects and Microorganisms

Chemical Name	Concentration[a]	Source of Chemical
Botulinum toxin	0.00000003 mg/kg	Bacteria (*Clostridium botulinum*)
Tetanus toxin	0.0000001 mg/kg	Bacteria (*Clostridium tetani*)
Diphtheria toxin	0.0003 mg/kg	Bacteria (*Corynebacterium diphtheriae*)
Black widow spider protein	0.55 mg/kg	Black widow spider (*Latrodectus* sp.)
Aflatoxin	1 mg/kg	Mold (*Aspergillus flavus*)

mg/kg = milligrams of chemical per kilogram of body weight.

[a] Amount needed to cause death from a single dose.

All types of organisms contain chemicals, some of which are toxic and dangerous. For example, we all know about the bite of a rattlesnake or black widow spider; their bites are poisonous because they cause specific toxic effects in our nervous systems. Therefore, it isn't surprising that there are natural ingredients in our foods that result in toxic effects if enough is eaten. In fact, these natural chemicals are widespread in our food supply. It may be surprising to learn that the great majority of highly toxic chemicals occur naturally in the environment, and are produced by animals and plants. These chemicals are produced by animals and plants primarily to serve as either defense mechanisms against predation (e.g., digitalis in plants protects against the plants from being eaten by animals), or as a mechanism to capture prey for food (e.g., rattlesnake venom paralyzes the prey so the snake can eat it).

Manufactured chemicals that we apply to our foods are also toxic at certain levels, but typically at much higher levels than those for the natural chemicals discussed above. These chemicals are known as food additives. Lead was an early food additive used to flavor Roman wines, as discussed in Chapter 1. Even as late as the mid-1900s, toxic chemicals such as thallium (a metallic element) were added to beer to serve as foaming agents in order for the poured beer to have a good head. This practice ceased after men began going bald from thallium exposure. Today, food additives are regulated in the United States by the Food and Drug Administration (FDA). The registration of food additives for use in specific foods is regimented, and is designed to ensure that the levels in food are safe for human consumption. The Delaney Clause, adopted in 1958 in the United States as part of the Food Additives Amendment, requires that any food additive be found "safe" before the FDA approves it for use in food. This means that no chemical can be used as a food additive if there is a known potential for it to cause cancer.

Historically, registration was less comprehensive and some additives were put in food that caused toxic effects. This was usually because the right type of test was not required to register the chemical, or the incidence of effects was so small that impacts were not seen until millions of people were exposed. Most often, this involved cancer tests. Scarlet red and butter yellow were historically used as coloring agents. In the 1970s it was found that these additives contained a type of chemical called a nitrosamine, which caused cancer. This discovery helped increase the types and numbers of tests necessary to register food additives to include cancer studies.

Natural products, including herbal products derived from natural products, are exempt from regulation as food additives by the FDA because the chemicals they contain occur naturally and have not been added by people. However, the majority of toxic chemicals we eat (including many that can cause cancer) occur naturally in our food. In fact, many toxic natural products are much more potent than any of the food additives that have ever been used. Yet media coverage of chemicals in food usually focuses on chemicals that have been added to food. The natural chemicals present in foods have often only made the news when an accidental poisoning has occurred, such as when someone dies from eating deathcap mushrooms (*Amanita phalloides*).

An example of a natural food that can cause toxicity is the puffer fish (also called the blowfish), which is a delicacy that is prepared by specially trained chefs and eaten in Japan. The puffer is a fish that lives primarily in tropical waters along coral reefs. It produces tetrodotoxin, one of the most potent neurotoxins known to humans, in its liver and ovaries. Less than 1 milligram of tetrodotoxin can kill a healthy adult; there is enough chemical present in one puffer to kill the person eating it, unless the glands that produce the toxin are carefully and completely removed from the fish prior to cooking. If the fish is not prepared properly and the chemical has been ingested, death occurs within about 2 hours from respiratory paralysis. An antidote can be given, but only up to about one hour after ingestion. After that, toxic effects are irreversible. About 100 cases of Fugu (the Japanese word for this fish) poisoning are reported every year. About half of these cases are fatal.

Another toxic effect that can result from improperly stored or prepared food is food poisoning. One type of food poisoning is due to the Botulinum toxin; toxicity associated with this chemical is called botulism. The Botulinum toxin comes from a bacteria (*Clostridium botulinum*) that can be present in improperly stored foods. These foods include canned products (green beans, corn, beets, asparagus, chili peppers, mushrooms, spinach, figs, olives, and tuna), smoked fish, fermented food (e.g., salmon eggs), and improperly cured hams. Botulinum is one of the most toxic chemicals to humans (see Tables 2.1 and 2.5). This toxin prevents transmission of neural signals, leading to muscle paralysis and death. The fatality rate for botulism ranges from 35 to 65 percent, and death occurs from 3 to 10 days following exposure. In nonfatal cases, botulism can last for 6 to 8 months, and symptoms involve trouble breathing.

The more common type of food poisoning is due to another, related bacteria (*Clostridium perfringens*). This bacteria can be present in raw and cooked foods, especially meats, poultry, gravy, stew, and meat pies. Some strains are resistant to cooking, meaning that even properly preparing the food might not kill the bacteria. At levels typically found associated with food poisoning, symptoms include gastroenteritis (e.g., nausea, vomiting, cramps). The effects of this bacteria typically last no more than one day.

A very different type of food toxicity involves toxins that are produced during the food preparation process. For example, charbroiled meats or smoked products (e.g., fish, sausage) typically contain polycyclic aromatic hydrocarbons (PAHs), which are formed during the cooking process (including barbecuing). Therefore, you will be exposed to small amounts of these chemicals when you eat meats prepared in this fashion. These are chemicals naturally produced through combustion. They are also formed from forest fires, and are considered products of incomplete combustion. PAHs are also present in diesel fuels and car exhaust, and in smoke released from chimneys.

Many PAHs are known to cause lung and skin cancer in animals, and might cause these effects in humans. But the dose from one meal is typically so small that we do not notice toxic effects after eating. Indeed, we might never exhibit toxic effects due to PAHs over our lifetime. However, the risk of developing lung or skin cancer might be slightly higher over the course of a lifetime from eating broiled or barbecued meats. This is a risk many of us are willing to take. See Chapters 8 and 9 for a more in-depth discussion of risks and decisions about which risks are acceptable.

Other examples regarding food toxicology involve fungi and plants. Wild mushrooms are often harvested and eaten with dinner in both the United States and Europe. One species of wild mushroom is the deathcap mushroom. This mushroom contains two very potent liver toxins, amanitin and phalloidin. Both of these chemicals are small protein-like chemicals that are preferentially taken up into the liver when eaten (see Chapter 6 for the basis of this preferential uptake). This species of mushroom is responsible for the majority of mushroom-related deaths across the world. The worst toxic effects are due to amanitin, which binds to specific parts of liver cells and prevents proteins from being synthesized. About three days after eating these mushrooms, liver damage begins to occur due to the lack of protein synthesis. The liver damage worsens with time. In cases involving low dose levels, the injury can be repaired because only a small part of the liver is affected. However, if the dose is high enough, the entire liver can be affected. Other than a liver transplant, there is no treatment for these toxic effects. This provides a good example of how a simple thing like picking mushrooms for dinner can lead to great risk of toxic effects, and even death, unless one is knowledgeable about differentiating between safe and toxic species.

Liver injury can also occur from using herbal remedies or teas containing comfrey (*Symphytum* sp.). Comfrey is a traditional European dietary and medicinal herb, and is also available in Australia and in the United States.

Comfrey and similar species (e.g., groundsel [*Senecio* sp.]) contain chemicals known as alkaloids. Alkaloids represent a wide range of structurally related chemicals that can differ widely in their relative toxicity. Specific alkaloids contained in comfrey are suspected to cause cancer in humans (e.g., echimidine, symphytine), and are known to cause liver toxicity other than cancer. Based on this information, the Canadian government took action in 1987 to prohibit the sale of comfrey for medicinal purposes. Due to the lack of regulatory control of natural products in the United States and other countries, comfrey is readily available in many markets. As with PAHs, the typical dose from using comfrey at any one time is low, and toxic effects are therefore unlikely to result. However, when used over a long period of time or at high levels, the risk of liver injury or cancer could be high.

If these same chemicals were manufactured and released into the environment, they would be regulated as carcinogens and their use restricted or banned. If they were manufactured, they could not be used in foods (as additives) because they could cause cancer. This underscores the differences between how natural and manufactured chemicals are regulated, and how they are viewed by the media and the general public. The toxicity of two chemicals might be identical, but the manufactured one will generate public outcry and regulatory response, while the natural one will be ignored. However, the chemicals may have similar impacts on our everyday lives and should be treated as equal problems and risks by the public.

Examples of Pollution Incidents

The examples discussed in the previous section illustrate how toxicology can affect us on a daily basis. Other toxicology-related events occur that impact how we regulate and manage chemicals. Many of these regulations and management decisions have resulted from specific pollution incidents. Some of these incidents have been one-time events, while others occur daily. We will look at some specific examples related to air and water pollution in the following sections. These examples are not meant as a comprehensive summary of pollution incidents across the globe. Instead, they serve to illustrate incidents that have helped shape policies regarding worldwide chemical pollution. These incidents have improved our understanding of the severity of the toxicity of certain chemicals in specific species.

In addition to incidents related to specific environmental media (i.e., air, water, sediment, and soil), catastrophic events occasionally occur that have impacts across all media—for example, the Chernobyl nuclear accident in the Soviet Union, atomic bomb testing, and the bombing of Hiroshima in World War II. Although the focus of this book is on non-radioactive chemicals, these examples are worth mentioning because of the magnitude of their effects and because they provide us with data related to the effects of chemicals on humans and wildlife species in the environment that cannot be duplicated in a laboratory. In these examples, we all serve as guinea pigs. It is hoped that we will use the data from these unfortunate incidents to better understand how radioactivity affects living organisms, and how we can minimize these effects in the future. In this way, toxicology can be used to increase our safety related to chemical exposure. Applying toxicology to everyday life in this manner is one of the best uses of the discipline.

Air Pollution

Probably the world's worst industrial disaster for nonradioactive chemicals occurred in Bhopal, India in December 1984. About 30 tons of methyl isocyanate was accidentally released from an industrial plant into the city's air. The lightweight chemical was released as vapor into the atmosphere. Methyl isocyanate is used as an intermediate in the synthesis of some pesticides. The shanty-like homes in the area had no air filtration system, and so vapors also entered homes. About 3,000 people were killed from the release of this chemical, and about 200,000 were injured or permanently disabled. Upon contact with the water vapor in the atmosphere, methyl isocyanate becomes a direct irritant, causing burns wherever it contacts the body. The deaths were mostly due to pulmonary edema (fluid in the lungs), which resulted from the direct irritation of the cells lining the airways in the lungs. Permanent disabilities included blindness from burning of the eyes. This terrible accident occurred over a matter of minutes during the night.

To put this incident into perspective, consider that, in the United States over a 7-year period straddling 1984, only 309 deaths were reported nationwide associated with chemical releases into all media combined. The Bhopal incident was unique both in the severity and the large number of people affected over a single release at one time. This event had a regulatory impact worldwide. In the United States, for example, this accident helped build support for a more stringent amendment to the Clean Air Act, first signed in 1970, that went into effect in 1990. This amendment listed 189 chemicals for which special standards and risk assessments were required by the end of that decade. These chemicals were selected by the U.S. EPA based on their relative toxicity and the volume of release. Methyl isocyanate was on that list.

More typically, chemical releases have an impact over a long period of time. An example is the release of dioxins into the air and soil at Seveso, Italy, which occurred in July 1976. The release was related to the production of trichlorophenol at a plant about 20 miles north of Milan. No explosion occurred. Instead, the release occurred more than 6 hours after a chemical reaction within a chemical reactor at the plant had been completed. The chemical reaction produced heat, which raised the temperature of the contents of the reactor. This rise in temperature led to additional, spontaneous chemical reactions that increased pressure within the reactor. The increased pressure ruptured a seal, which led to release of the reactor contents into the atmosphere. The resulting toxic cloud of vapors released from the plant contaminated several thousand acres of a densely populated area of Seveso. Vegetation, birds, and animals near the plant were impacted within days of the release. Many herbivorous animals (e.g., rabbits, sheep) died from eating contaminated plants. Nine days after the release, dioxins were found to be present in the plants, animals, and soils of the area. Skin lesions were reported by residents, especially in children that had more direct contact with the contaminated soils than adults. Two weeks later, 179 people were evacuated from the area immediately surrounding the plant. After additional sampling showed the extent of dioxin contamination was more widespread than originally believed, about 550 more people were evacuated from a larger area around the plant.

Overall, the accident was considered directly responsible for the deaths of about 3,300 small animals, and about a dozen domestic animals. Over the next two years, cleanup activities were conducted. Soil was excavated and removed from the contaminated zones, and some homes were destroyed due to contamination. Based on uptake of dioxin from soil and food, an additional 77,000 animals were killed over this time period to prevent them from spreading contamination in predators eating the wild animals and humans eating the domestic animals.

After this two-year period, the majority of evacuated residents were allowed back into their homes. Based on medical records, no human deaths were attributed to the release. However, 15 people had some scarring from severe skin lesions due to contact with the dioxin. An additional 183 people fully recovered from skin lesions. The skin lesions caused by dioxin are known as chloracne.

Compared to the Bhopal incident, the degree of toxic impact from this accident was low. However, when the release occurred, our toxicology knowledge related to dioxins was limited to animals or with humans exposed to very high concentrations in enclosed spaces. Because of this lack of knowledge, experts were not sure what the long-term effects might be. As a result, the media reported the potential worst effects, which may have helped perpetuate the fear and uncertainty surrounding the accident.

This incident, and follow-up research on toxicity over time resulting from the release, led to the tremendous precautions taken related to the Times Beach, Missouri and Love Canal, New York removal efforts of dioxin found in soil. You may remember the pictures of people walking around in "moon suits" excavating soil from these areas. More recently, after the U.S. EPA more fully evaluated dioxin toxicology in humans, they admitted that they took extreme measures that were likely not necessary to deal with these situations.

An example that occurred in the United States relates to workers involved with manufacturing asbestos. In the late 1960s, asbestos use soared based on its insulation properties. Uses during this time included components of rocket engines used in the space program, paper and cement products, and components of gaskets. Asbestos is naturally found in certain rock types (e.g., serpentinite), and is extracted from the rock and manufactured in indoor factories. Lots of dust containing asbestos was released into the indoor air environment during the process. Several years after people starting working in the factories, several cases of a new type of lung cancer called mesothelioma began appearing in these workers. Soon it was determined that the asbestos fibers released into the air and inhaled caused this type of lung cancer.

These incidents led to the current restricted use of asbestos by the U.S. EPA. This restriction is the reason why asbestos is being removed from many older buildings. Due to these regulations and the restrictions on asbestos production and use, asbestos-related cancers were estimated to account for only about 5 percent of the approximately 2,000 lung cancers from airborne chemicals (other than cigarette smoke) reported in 1993 (equal to about 100 cases). This is a dramatic reduction from the rates seen in the 1970s and early 1980s. Because many years might elapse between asbestos exposure and cancer, most of these cases likely are related to when asbestos was still manufactured, rather than recent exposure cases. Asbestos-related toxicity is still prevalent in other countries where asbestos is mined, manufactured, and used without regulatory restrictions (e.g., South Africa).

Two other examples of air pollution are diesel engine exhaust and secondhand smoke. All combustion engines release exhaust from the fuel used to power the engines. Three of the primary fuels used to power the majority of engines are gasoline, diesel, and jet A fuel. The exhaust of diesel engines was recently determined to be a human carcinogen by the U.S. EPA, and is also considered a human carcinogen by the International Agency for Research on Cancer (IARC). This is not surprising since diesel exhaust contains PAHs. As discussed in the food toxicology section earlier in this chapter, several PAHs are known animal carcinogens. This is likely to have dramatic implications for the future of diesel engines. Regulations might change to limit the release of diesel exhaust because of the lung cancer risk inherent in burning the fuel. There have as yet been no estimates as to the number of lung cancers due to diesel exhaust each year, but it could be as much as 5 percent of the total (100 cases).

Secondhand cigarette smoke is considered to be a human carcinogen by many of the scientific and regulatory bodies across the world, including the IARC and U.S. EPA. As previously mentioned, about 2,000 cases of lung cancer were estimated to be due to airborne chemicals in 1993. Secondhand smoke was not included in this total, but estimates for this same year indicate that an additional 2,000 cases of lung cancer were due to secondhand smoke in the United States that year. This is equal to the total number of lung cancer cases due to all other chemicals combined. Based on this information, the state of California has adopted strict rules regarding indoor air quality; smoking is not allowed in most public buildings, including restaurants.

All of these examples are directly linked to specific actions taken by regulatory bodies in response to toxicology information obtained for specific chemicals and the ways in which they are released. Because these regulations impact each of us, toxicology affects our everyday lives.

Growth was the focus of the beginning of the industrial age. This growth led to increased air pollution from industrial facilities and automobiles, impacting human health. The effect on health, including those described above, led to controls on air pollution. Due to these regulations, outdoor air quality has improved in the United States and parts of western Europe since the 1970s. It is encouraging to see that the regulatory changes adopted by governments are working to reduce air pollution in many locations. This improves our quality of life and decreases our risk of toxic effects from exposure to these chemicals. It is unfortunate that these changes and improvements usually only occur after people die or are injured. In most cases, this is due to lack of knowledge rather than poor management.

Water Pollution

Similarly to air pollution, waste discharge into surface water has increased during the industrial age. This led to incidents of pollution across the globe until regulations were put into place to limit the types and amounts of chemicals that can be released into surface waters from industrial processes. Two examples from Japan are presented to illustrate some typical water pollution incidents that arose from industrial waste discharge.

The elemental form of mercury was used as a catalyst to make acetaldehyde at an industrial plant near Minamata Bay in Japan. Acetaldehyde is used in the manufacturing of perfumes, plastics, synthetic rubber, and other products. In the 1950s some people developed a specific set of neurotoxic symptoms (effects on the nervous system). Effects included speech impairment, narrowing of the visual field, and ataxia (loss of coordination). In 1963 the disease was identified as organomercurial poisoning, and later became known as Minamata's Disease.

Between 1965 and 1974, 520 patients in Nigata were treated for this disease. All of these cases were caused by the mercury released from the plant. The cases were unexpected because the elemental form of mercury does not cause these effects.

However, once the mercury was released into the water, the aquatic plant life and microbes in the sediment of the bay converted the mercury, as part of their regular physiological functions, to an organic form known as methylmercury. The methylmercury was then taken up by fish and shellfish that ate the plants or animals that lived in the sediments. Local fishermen then ate the fish and shellfish, and it was these people that reported the toxic effects. This organic derivative of mercury can enter the nervous system while the original form cannot.

This complex series of events made it difficult to trace the toxicity back to the original release of elemental mercury from the plant. The research took several years. Today we know that the effects can occur in unborn fetuses, and this leads to neural effects during childhood. It is this effect of mercury that causes the most concern, and is the basis for regulations regarding levels of methylmercury in fish and shellfish that are allowed to be eaten in the United States and elsewhere. Partially based on this incident, fish and shellfish containing more than a certain level of methylmercury cannot be consumed because of the risk of adverse effects. In the United States, regulations are most often in the form of fish advisories for sport fishing.

A second incident also involved an element, this time cadmium. In the 1960s, mining wastes were dumped into rice paddies in Japan. The chemicals in the mining wastes dissolved into the water in the rice paddies, and also were absorbed into the rice itself. Middle-aged women that had calcium-deficient diets and multiple pregnancies began developing a set of symptoms that included extreme bone pain; lumbago; pain in the back, shoulders, and joints; a waddling gait; frequent bone fractures; and elevated levels of calcium and protein in their urine. This set of symptoms was called itai-itai (ouch-ouch disease) because of the pain associated with walking. It was later learned that the causative agent in this disease was cadmium that was present in the mining wastes. The middle-aged women that developed this disease had a daily cadmium intake about 200 times the typical intake in an unpolluted area. This incident led to the current regulatory limit of cadmium in water and food set by the U.S. EPA.

Another incident occurred in the United States, and also involved an element (lead). In Chapter 1 we discussed the neurological effects of lead on adults. It is also known that lead can cause neurological damage in children, including impacting intelligence. Across the county, elevated levels of lead were found in drinking water, not in reservoirs but at the tap in individual homes and schools. This was more prevalent in areas of low-income housing. These elevated levels of lead were traced to the piping used to bring drinking water into the homes or schools. Historically, water pipes contained lead. Nobody expected that lead, which barely dissolves at all in water, would be removed from the pipes in a process called leaching, and end up in the water we drink. After the link between elevated lead levels in tap water and lead in pipes was made, water pipes were replaced with new pipes that do not contain lead. This process of replacing water pipes is still in progress across the country.

Another example of water pollution involves species other than humans. After all, it is aquatic species that usually are first exposed to chemicals released into surface water (e.g., lakes, rivers, and oceans), and they are likely to be exposed over a long period of time. Probably the most infamous incident from the recent past is the large oil spill from an Exxon oil tanker in Prince William Sound, Alaska.

In this case, rather than chemicals causing toxic effects after being absorbed into the body, the toxicity was largely due to the thick oil covering the wings of birds and fur of marine mammals (e.g., sea otters). This oil prevented birds from flying and prevented birds and mammals from regulating their body temperature. The oil entered the intertidal zones and was distributed across rocks, on beaches, and in the sediments. Because of the wide distribution of the oil, essentially all aquatic organisms, ranging across the food chain from algae to whales, were affected to some degree. The ecosystem in the area was greatly impaired by this spill. Cleanup efforts ran into the billions of dollars as crews attempted to collect the raw oil.

Although the devastation was severe, certain bacteria actually can eat petroleum products like oil and use it as energy. The bacteria in the water actually provided a mechanism for remediation. Thanks to the efforts of the cleanup crews and the bacteria, the ecosystem has mostly recovered and populations of most species are returning to near normal levels. A regulation known as the National Contingency Plan, which primarily deals with oil spills, was established by the United States in 1990, prior to this spill, because other oil spills that had occurred around the world had similar effects. This regulation led to the largest fine ever levied by the United States for an environmental release of oil into water—more than $5 billion. Exxon paid about $3 billion dollars in cleanup costs.

Media Reporting of Toxicology

We have already discussed the media's focus in reporting issues related to toxicology (e.g., food additives versus natural products, in the Food Toxicology section at the beginning of this chapter). However, it may be helpful to expand on this discussion because the media (e.g., newspapers, radio, and television) are the public's primary source of information relative to toxicology. Several factors may help prevent reporters from presenting enough information to give the public the full story:

- "Bad" news is more dramatic.
- Reporters usually lack scientific qualifications.
- Stories must be turned in on a deadline, whether or not all sources were reached to present a balanced view.
- Biased sources can lead to misreporting.

We will discuss this further in Chapter 9, specifically regarding risks of toxic effects related to chemical exposure. The three examples below serve to illustrate the benefits, false alarms, and inflammatory and biased reporting, respectively, that may occur in the media's coverage of toxicology.

The insecticide DDT was first produced in the 1920s, and was extensively used between 1945 and 1965 to control insects. DDT was an excellent insecticide because it was very effective at killing a wide variety of insects at low levels. However, the chemical properties that made this such a good insecticide also made it persist in the environment for a long time. This persistence led to accumulation of the pesticide in non-target species, especially raptorial birds (e.g., falcons). Due to the properties of DDT, the concentration of DDT in birds could be much higher than concentrations in insects or soil. Birds at the top of the food chain (e.g., pelicans, falcons, eagles, and grebes) had the highest concentrations of DDT. Although the amount of DDT did not kill the birds, it interfered with calcium metabolism, which led to thin eggshells.

As a result, eggs would crack during development, allowing bacteria to enter, which killed the developing embryos. This had a great impact on the population levels of these birds. Peregrine falcons and brown pelicans were placed on the endangered species list in the United States, partially due to declining reproductive success of the birds from DDT exposure. Rachel Carson, a journalist, published *Silent Spring* in 1962, helping to draw attention to this problem. This was the very beginning of the environmental movement in the United States, and is an excellent example of reporting by someone affiliated with the media that identified a problem and warned of many similar problems that could occur unless restrictions were put in place related to chemical pesticide use. Partially as a result of this book, the link between DDT and eggshell thinning was documented by scientists. This led to DDT being banned for all uses in the United States in the early 1970s. DDT is also banned in Europe.

Although it is an excellent insecticide, the known effects on other species, such as raptors, were not acceptable for continued use in the United States and Europe because other alternatives were available. However, DDT is still used in developing countries because it is inexpensive and highly effective. Other alternatives are too expensive for these other countries to use.

An example of media reporting that was unnecessarily alarming to the public is the reporting of the use of the herbicide daminozide (Alar) on apples. Daminozide is a growth regulator that was used to control the vegetative and reproductive growth of orchard crops such as apples, cherries, nectarines, peaches, prunes, and pears in the late 1980s. It was also used on ornamental plants to control the size and shape of the stems.

Early in 1989, the media reported that Alar was a human carcinogen, was present at detectable levels in apples sold at supermarkets, and would lead to an unacceptable risk if people ate apples containing Alar. Actually, a breakdown product of daminozide that forms in water (1,1-dimethylhydrazine)—not the chemical itself—has been shown to cause cancer in laboratory animals. In 1986, a fact sheet was produced indicating that there were no restrictions on the use of Alar for registered food crops such as those listed above. However, based partly on public concerns from media reports, the U.S. EPA officially banned Alar from use on any food crops in 1992. Although the decision to ban Alar from use on food crops was probably the correct decision, the manner in which it was reached was based on public perception from media reports.

The issues not accurately reported by the media included: (1) Alar itself was not a carcinogen, and it was not reported whether the carcinogenic breakdown product was present on the apples sold in supermarkets, and (2) some levels of Alar on apples would lead to exposures so low that the risks would be below any levels of concern. This is why U.S. EPA approved daminozide use on apples in the first place.

By not fully reporting these issues, thousands of people became very concerned about getting cancer from eating apples. The risk of developing cancer from eating apples contaminated with the carcinogenic breakdown product of Alar was very low compared to other cancer risks. We will discuss this issue of relative risks more in Chapter 9. In addition, the media reporting led to concern that people would experience toxic effects in the short-term (e.g. acute effects), even though this concern was unfounded. For example, using the ranking scale shown on Table 2.1, Alar would be considered "practically nontoxic," much like table salt. This illustrates the importance of understanding media reports so that the actual problems can be understood.

The final example discussed here concerns the fuel oxygenate MTBE (methyl-t-butyl ether). The use of fuel oxygenates was mandated by Congress in the early 1990s to reduce air emissions of gasoline-related toxic chemicals (e.g., benzene, a known human carcinogen). One of these fuel oxygenates, developed for use in gasoline by an oil company, is MTBE. Prior to its use in gasoline, MTBE was commonly used in chemistry laboratories as a solvent in chromatography. This chemical has a noxious odor and taste, and can be smelled and tasted at very low levels in water (e.g., 10 parts MTBE per billion parts of water). It dissolves easily in water, and gasoline leaks from underground storage tanks at gas stations has led to the chemical reaching groundwater. Some of this groundwater, especially in southern California where surface water is scarce, is used as drinking water.

Because of the odor and taste of the groundwater due to MTBE, residents in southern California were supplied with bottled water. MTBE is not a particularly toxic chemical—it would be categorized as "practically nontoxic" using the scale shown on Table 2.1. However, one report from a laboratory in Italy indicated it could cause kidney tumors in laboratory rodents. After much additional research in the United States failed to confirm the Italian study, it was recently concluded (by state of California toxicologists) that MTBE is not a carcinogen. U.S. EPA considers MTBE to be a possible human carcinogen, but research is ongoing.

The media has typically reported that MTBE is a carcinogen, and many environmental groups have requested the additive be banned from use in gasoline because of its impact on our groundwater and surface water. It is likely that MTBE will be phased out of use in California with the support of environmental groups and politicians. The media has played a large role in generating support for such decisions.

People typically associate bad odors with toxicity. As a result, it is easy to understand how the reporting of MTBE has become skewed. For example, a state of Maine regulator said that "public concerns over MTBE lead to all levels of MTBE posing a great health threat to the public, regardless of its toxicity." This statement literally means that public concern over a chemical can, by itself, make that chemical a health threat, even in the absence of an actual health risk. Clearly this cannot be true. It is unclear if the regulator actually meant what this implies, but statements like these, which violate the most basic principles of toxicology (e.g., implying that a health threat is unrelated to toxicity), add to the public's confusion over the real issues. Hopefully this book will help you understand these issues, and which comments and reports should be dismissed based on their purely emotional or political bases.

Limitations of Our Current Knowledge

There are several hundred thousand chemicals either naturally produced or manufactured that have some use for humans. We have adequate human toxicology information for less than 100 of these, and adequate animal toxicology information for less than 1,000. Therefore, we do not have adequate information for almost all chemicals. As we gain knowledge about chemicals, we will find that some present greater concerns than we first thought (e.g., DDT) while others will not pose the threats we thought they might (e.g., MTBE). Lack of chemical knowledge is typically handled by regulatory agencies in a conservative way. In other words, if we do not know the actual toxicity of a chemical to humans, we assume it is high (e.g., precautions taken regarding dioxin at Times Beach and Love Canal). This leads to regulations that should protect us from toxic effects in most circumstances.

As we discussed in this chapter, natural chemicals in our food can lead to toxic effects, as can manufactured chemicals. We carefully regulate manufactured chemicals that are in food or are released into the environment. We do not similarly regulate natural products. This means that there is the potential for toxic effects to result from exposure to just about any chemical or product if exposure is high enough or the chemical or product is misused. This book will hopefully enable to you identify when a threat is real, when it can be ignored, or when a small risk might be worth taking. Nevertheless, the number of real threats to our health from manufactured chemicals is very low.

Additional Reading

Carson, R., 1962. *Silent Spring*. Houghton Mifflin Company, Boston, Massachusetts.

3

Essential Concepts of Toxicology

As we have seen, all chemicals can be considered toxic. However, unless we are exposed to a chemical, we will not have toxic effects from it. This fact is what allows us to live our lives without being impacted by the majority of these chemicals. This introduces one of the most important concepts in toxicology, which can be summarized by the following simple relationship:

Toxic Effects = Potency x Exposure

A toxic effect is an adverse response in an organism caused by a chemical. Potency generally describes the rate at which a chemical causes effects. More potent chemicals have higher rates than those that are less potent. For example, salt has low potency because there is a wide range of amounts over which the degree of toxic effects, if any, changes very slowly.

On the other hand, one drop of strychnine, a natural chemical from the seeds of *Nux vomica* that attacks our nervous system, might kill us. For this chemical, very little change in amount is needed to have toxic effects range from nothing to death. This is a very potent chemical.

The potency of chemicals varies widely, but is never zero. Why then are we not constantly at risk from chemicals? Because in order for toxic effects to occur, we must be exposed to these chemicals. Exposure can be controlled so that it can be zero. If exposure is zero, there are no toxic effects because the product of potency times exposure is zero.

This chapter introduces the reader to this and other essential concepts of toxicology. The concepts to be discussed in this chapter include the following:

- Exposure
- Mechanisms of toxic action
- Species variability
- Dose-response (i.e., potency)

A basic understanding of these concepts allows us to understand how to interpret the above relationship under many different conditions and for many species, including humans and wildlife. Once these concepts have been introduced, then specific types of toxic effects can be discussed with the proper perspective, as is done in Chapter 4.

Each of these four concepts can itself be the focus of an entire book. The goal of this chapter is to introduce the reader to how these concepts govern the majority of the science of toxicology, and applications such as risk assessment.

Exposure

Typically, toxicology focuses on how chemicals exert toxicity in living systems. Toxicity can be defined as the intrinsic degree to which a chemical causes adverse effects. Evaluating this issue is usually done by administering certain amounts of a chemical to a target species and evaluating the response, such as death or reproductive impairment. It may seem surprising then to see that the first concept discussed here does not even deal directly with toxicity. Rather, it focuses on exposure. As discussed earlier, without exposure, no toxic effects would occur.

The majority of environmental and workplace regulations and training focus on minimizing exposure to chemicals. For example, the Clean Water Act and Clean Air Act target allowable concentrations that can be released into our environment. These targets vary for different chemicals. This is intended to account for differences in the potency of chemicals, and is designed to limit possible exposures to levels below which toxic effects are expected. Workplace safety requirements in places where chemicals are handled usually include eye protection and gloves. Chemicals that readily evaporate (or volatilize) into air are typically required to be used in hoods that direct vapors up and away from people using the chemicals. These requirements are all designed to minimize or eliminate human contact with a chemical. All of the labeling instructions found on pesticides and other chemicals sold for home use are similarly designed. These requirements and instructions do not lower the toxicity of a chemical; they lower the exposure potential to levels below which toxic effects are expected.

Exposure Routes

Chemical exposure is defined as contact with a chemical by an organism. For humans, there are three primary routes of chemical exposure:

- Through breathing (i.e., inhalation)
- By eating (i.e., oral)
- By touching (i.e., dermal contact)

Another minor route of exposure includes through the eye (ocular contact). Each of the three main exposure routes and examples are discussed below.

Inhalation Exposure

This exposure route involves breathing chemicals either as vapors (e.g., toluene in glue; formaldehyde in new carpets) or particulates (e.g., smog). Inhaled chemicals will pass through the nose and mouth into the respiratory tract, which includes the trachea, bronchi, bronchioles, and alveoli. The bronchi, bronchioles, and alveoli comprise the lungs.

The size of the airways decreases from the trachea to the alveoli. This decreasing size, along with the many angles formed by branching airways, prevents many particles containing chemicals from reaching the alveoli. Oxygen and carbon dioxide are exchanged in the blood across the alveoli. However, chemicals in gas form (i.e., vapors) will not lodge in the airways and can reach the alveoli, along with the oxygen taken in through breathing. These vapors can then enter the bloodstream and be transported to organs where toxicity is manifested. Carbon monoxide, toluene, and formaldehyde are examples of vapors that can pass across the alveoli into the lungs through inhalation. Carbon monoxide replaces oxygen in red blood cells, leading to asphyxiation. Toluene enters the central nervous system and causes anesthetic effects. Formaldehyde can impact respiration, especially in asthmatic children. It has also been shown to cause nose tumors in laboratory mice.

Many particulates are too large to reach the alveoli; they will be deposited in larger airways (e.g., bronchi). This includes the particulate matter components of smog (the vapor components of smog are not similarly affected). Only particles smaller than about 10 microns in diameter are likely to reach the alveoli. One micron is 0.000001 of a meter. Our airways are lined with small hairs, called cilia, which beat up to the mouth (i.e., the opposite direction of inhaled air). The purpose of these cilia is to move deposited particulates up away from the alveoli and towards the mouth. These particles can then be ingested, where they can be eliminated from the body or absorbed from the gut. Very small particles may reach the alveoli and pass into the blood.

This demonstrates that we are physiologically adapted to remove contaminants from our air passages before they reach the lung and bloodstream. The hairs in our nose are our first defense against particulates; cilia represent a secondary defense mechanism. For people with impaired respiratory systems (e.g., smokers, asthmatics), the cilia may be damaged or covered with mucus, preventing them from removing particles before they reach the alveoli. Therefore, the risk from inhalation exposure will be increased if the airways are damaged.

Even though we have these defense mechanisms, many chemicals exert their toxic effects directly on the lungs (e.g., paraquat). These chemicals are the ones from which we are most at risk from inhalation exposure. Paraquat is an herbicide that was historically used to eliminate pests from plants, including marijuana plants. Paraquat sprayed on plants that are then smoked is directly transported to the lungs. In the lungs, paraquat binds to fatty molecules in cell membranes of the lungs, which leads to the replacement of lung tissue with fibrous tissue that is not elastic. Therefore, the lungs fail to expand and contract, and death from respiratory failure ensues. This effect takes only a few days to occur, and there is no cure for this type of exposure. As little as one teaspoon can cause this effect. This illustrates that inhalation exposure, depending on the chemical and the amount, can lead to very serious toxic effects.

Oral Exposure

This route of exposure includes ingesting food, water, and other substances (e.g., dirt) that might contain chemicals. The Food Toxicology section in Chapter 2 focused on ingestion exposure.

Ingesting chemicals does not automatically mean that some of the chemical will enter your body. The digestive tract is a long tube that starts at the mouth and ends at the anus. It is not until a chemical is absorbed through the intestines that the chemical enters the body. To use a gross example, if your intestines were unwound and the gastrointestinal tract was made into a long tube, one could look into the mouth and see daylight out the other end. Therefore, unlike the lung, oral exposure does not directly lead to chemicals entering the body.

This illustrates the difference between exposure, which is contact with a chemical, and dose, which is the amount of a chemical that enters the body. Using tuna fish as an example, ingesting mercury in tuna is an exposure to mercury. The amount of mercury that is absorbed across the intestines and enters the bloodstream is the dose of mercury. The dose controls whether or not toxic effects will result from exposure.

Although eating food and drinking water are the most common forms of oral exposure, others are also important. One common example used in the environmental field is incidentally ingesting soil contaminated with chemicals.

Most of us do not intentionally ingest soil, but we will contact soil through some regular activities (e.g., landscaping, gardening). This soil remains on the skin for some period of time. If someone fails to wash their hands and then puts their hand in their mouth, the soil particles on the hand can enter the mouth and be ingested. Chemicals sticking to these soil particles will also be ingested. This can also result from someone eating a sandwich on a lunch break without washing his or her hands. These particles can be too small to see with the naked eye.

Surprisingly, this is the type of exposure that typically leads to the greatest dose and risk of toxic effects at sites contaminated with chemicals. In part, this is because the intestines are designed to absorb chemicals (e.g., nutrients). Unwanted chemicals will get absorbed along with the nutrients, so absorption can be high for some chemicals. This route of exposure can lead to very high doses because concentrations of chemicals are generally much higher in soil than in other media (e.g., air). Soil used for landscaping by homeowners is regulated to prevent excessive amounts of chemicals from being present.

Dermal Exposure

This pathway of exposure occurs when a chemical touches the skin. Similar to the oral route of exposure, contact with the skin does not directly lead to a dose. The chemical must move through the skin, the outer layers of which are dead tissue, to the bloodstream before it enters the body. Because the outer layers of the skin are dead, they represent a barrier against absorption. In order for a chemical to be absorbed across the skin, the chemical must be dissolved in fluid. The fluid can diffuse through the skin, moving chemicals with it, until living tissue is reached. At that point, absorption occurs.

One common example of a chemical to which we are primarily exposed via the dermal route is DEET (N-N-diethyl-m-toluamide), which is used in mosquito repellants. Because DEET is typically in a gel or liquid form, some of the applied chemical can cross the skin and be absorbed into the body.

Another example is formalin used as a preservative for biological tissues in school labs. Formalin is a solution containing about one-third formaldehyde, which can cause many different toxic effects, depending on the amount of exposure. Preserved animals are often stored in formalin until distributed to students for dissections, etc. Dermal exposure to formalin will then occur, primarily through the hands. The instructors will generally have a higher exposure because they directly contact the formalin solution. Formalin dermal exposure will lead to very dry skin, but typical exposures in school lab settings should not lead to toxic effects. Wearing gloves will (basically) eliminate this exposure because the formalin will not contact the skin through gloves. This indicates that the dermal exposure route can be controlled through the use of protective clothing. This is the basis of many health and safety requirements in the workplace.

Exposure Duration

The amount of exposure to a chemical depends in part on the exposure route. In addition, the length of time over which exposure occurs will affect the total amount to which someone might be exposed. Duration is classified as:

- Acute
- Subchronic
- Chronic

Acute Duration

Acute exposure involves single episodes, or otherwise very short periods, such as a week or less. The Bhopal air pollution incident discussed in Chapter 2 involved an acute exposure duration. The releases occurred, people inhaled the chemicals, and the chemicals dispersed in the atmosphere or settled to the ground. These types of durations typically only lead to toxic effects if exposure occurs to relatively high concentrations of a chemical. An example of an acute duration would be exposure to poison oak or poison ivy while hiking. The exposure is essentially instantaneous, although the effects may last for a week or more if someone is allergic to the toxins in the plant. Spills requiring emergency responses also involve acute durations. The spill occurs, is contained, and is cleaned up in a matter of hours. Those near the spill may have acute exposures to the spilled chemicals. The emergency-response team, however, might have acute and subchronic or chronic exposure because they may deal with spills involving this chemical over several years (see below). Therefore, the duration of exposure can differ for different people exposed to the same chemicals under the same conditions.

Subchronic Duration

This represents a length of time between acute and chronic durations. This duration can extend from more than a week to less than 7 years for humans. Students in chemistry lab will have subchronic exposure over their college years to typical solvents used in the labs (e.g., methanol, chloroform). A typical emergency-response technician working at the same job for 5 years could have subchronic exposure durations to commonly spilled chemicals (e.g., gasoline).

Chronic Duration

These are the long-term exposure durations that typically lead to the highest overall exposures. A chronic exposure duration is typically defined as one spanning at least 10 percent of a lifetime. For humans, this is considered seven or more years. Although daily exposure might be low, the total exposure over a lifetime might be very much greater than higher exposures over shorter time periods. There are numerous examples for this type of duration. Someone who does dry cleaning for a living will have chronic exposures to trichloroethylene (TCE), a cleaning solvent that has anesthetic effects and might cause cancer. We all receive chronic exposures to gasoline vapors during refueling, and to smog through breathing. Smokers are chronically exposed to nicotine, cadmium, and other toxic chemicals contained in the tobacco and additives to cigarettes. Casino workers may have chronic exposure to secondhand smoke from those patrons smoking at their tables. Careers in the chemical or environmental industry can lead to occupational exposure to chemicals over a career. Someone living directly downwind of a smelter for many years will have chronic exposures to released chemicals from the stacks.

In toxicology and risk assessment, chronic exposure is of primary interest because these types of exposures are the ones that are most relevant for the typical person. However, as will be seen later (Chapter 5), relevant toxicology data for chronic exposures is often lacking.

Exposure Frequency

The duration of exposure alone does not indicate the amount of chemical someone might have been exposed to. In addition to duration, the frequency of exposure over that time period will impact the level of exposure. Chronic exposure may be insignificant if someone is only exposed for five days each year. As discussed earlier, we are exposed to smog daily over a chronic time period. Although eating tuna containing mercury may also be of chronic duration, the frequency at which we eat tuna is much less than our exposure to smog. Therefore, our overall exposure to mercury from tuna fish should be much less than our exposure to chemicals in smog.

Recently, much interest has been focused on the toxicity of diesel exhaust, especially as a possible cancer-causing mixture. Does this mean that we should avoid following behind buses, diesel trucks, or cars? Likely our activity patterns lead to only infrequent exposures to diesel exhaust, even though the exposure may persist over many years. Therefore, it is unlikely that these very brief exposure episodes, on the order of a minute or two daily, will lead to toxic effects.

The combination of exposure routes, duration, and frequency will govern the amount of chemical to which someone might be exposed, which will have an impact on the likelihood of receiving toxic levels. This can be expressed by the following relationship:

Exposure level = (sum of exposure routes) x exposure duration x exposure frequency

It is important to note that these exposure levels do not equal doses, which govern the toxic responses from chemical exposure. It is not until the chemicals are actually absorbed into the body that someone receives a chemical dose. This concept is more fully discussed in Chapter 6.

Mechanisms of Toxic Action

Now that we have discussed the types of exposures by which we may contact chemicals, this section introduces the concept of different types of toxic actions by chemicals. In general, there are two distinctly different categories of toxic action: chemicals that cause cancer and those that do not. Within each category, there are many specific types of toxic action. These types of actions will be discussed in Chapter 4.

Cancer Mechanisms

Currently in the United States, one in three people develop some form of cancer, or uncontrolled cell growth, in their lifetimes. Some of these cancers are typically fatal (e.g., lung cancer from cigarette smoking), while some are treatable and not usually fatal (e.g., skin cancer from exposure to ultraviolet light). This high incidence rate indicates that there are a myriad of cancer agents, some of which are chemicals. The biological mechanisms of cancer are described further below.

Cancer was first reported by Sir Percival Pott in the late 18th century, who noted that chimney sweeps were developing cancer of the scrotum. It wasn't until 150 years later that the chemical cause and mechanism was understood. One of the chemicals in ash, benzo(a)pyrene, was found to be responsible for the cancer. (This same chemical is also present in diesel fuel, and it and other related chemicals are formed when we burn meats on a barbecue. Yet the number of scrotal cancer cases is very small relative to the number of men that eat barbecued meats, implying that a certain dose is needed before such cancers develop.) The large time lag between identifying cancer and learning the cause underscores the complexity of cancer. One of the problems in identifying the causes of cancer is that cancer usually only develops many years after exposure to a chemical or set of chemicals.

We are learning more about various cancers, as well as their causes and potential cures, on a daily basis. Based on carefully controlled studies with laboratory animals and epidemiological (i.e., human) information (e.g., chronic workplace exposure), chemicals have been classified into those that are known human carcinogens, probable carcinogens, possible carcinogens, and those that are not carcinogenic. Very few chemicals are known human carcinogens, but many are considered probable carcinogens. This is because it is very difficult to be sure that a cancer that develops in humans is due only to chemical exposure. We all have different behaviors (e.g., smoking, drinking) that may contribute to cancer development. However, for certain chemicals, the type of cancer that results is very specific (e.g., mesothelioma of the lung from asbestos) or consistent (leukemia from benzene exposure). In these situations, we can feel more confident in directly linking exposure and toxicity.

Although there are many types of cancers and some types do not conform with the typical development of cancer, in general, cancer is the result of a series of events that results in uncontrolled cell growth. This process consists of four basic steps:

- Initiation
- Promotion
- Progression
- Cancer

Initiation is the mutation of a gene. When the initiator is a chemical, this typically occurs when the chemical interacts with and alters DNA. DNA is deoxyribonucleic acid, which contains the genetic code for all organisms. This genetic code is comprised of a series of genes, each of which governs specific actions (e.g., protein synthesis) that define and distinguish each organism. The altered DNA is then replicated, leading to the mutation.

These mutations are within the cell, and usually impact how cell processes work. Mutations that cause changes to our appearance (e.g., cleft palate) are different than those that lead to cancer. Mutations at the cellular level occur constantly in our bodies. The vast majority of these mutations either result in death of the cell or are repaired through our metabolic processes. It is only when the mutation allows the cell to survive, escapes repair processes, and successfully replicates during cell division that initiation is considered to have occurred. This step is irreversible. This does not mean the cancer will then develop; it means that if other events occur that allow these replicated, mutated cells to further multiply, cancer may develop.

Promotion is the process whereby the mutant cells formed during initiation are transformed into cancerous cells. This can occur by stimulation of cell division or other processes. This stimulation can be caused by chemicals, or by specific genes within our cells that mistake the mutated cell for a normal cell. Cell processes can reverse this promotion up to a certain point. The end process of promotion is the presence of numerous cancerous cells that are stimulated to divide and grow.

Progression involves multiple events over time that result in development of a tumor. It is at this point that a diagnosis of cancer may first be identified. This can take up to 30 years after initial exposure to a cancer-causing agent. Given this process and the fact that multiple events are necessary for cancer to develop, it is not surprising that we still do not fully understand all of its causes and find it difficult to ascribe a specific type of cancer to a single chemical. However, some chemicals are known to directly cause cancer, indicating that they are able to cause all steps in the progression of cancer from a normal cell (e.g., benzo(a)pyrene).

Because of this lack of knowledge about the majority of chemicals and their ability to directly cause cancer, regulatory programs such as those within U.S. EPA assume that one molecule can lead to cancer. This is a non-threshold approach to cancer. That is, exposure by one cell to one molecule of a chemical can cause initiation, which can then lead to promotion, progression, and tumor development. As we have seen from the above discussion, our cells have repair mechanisms that decrease the chances that one molecule can directly lead to cancer. Nonetheless, this assumption that there is no threshold for development of cancer has driven many regulatory programs for many years (see Chapter 8 for further discussion of this issue). Recently this assumption has been shown to be incorrect for some chemicals (see Chapter 10). We will discuss specific types of cancer and chemicals in Chapter 4.

Non-Cancer Mechanisms

Non-cancer toxicity includes effects that injure specific or multiple organs or systems (e.g., alcohol and the liver, lead and the nervous system). The mechanisms are almost as numerous as the number of chemicals. Specific effects and examples of chemicals that cause them are discussed in Chapter 4. The main point to make here is that, for noncancer effects, a threshold level is assumed. Below this threshold level, no toxic effects are expected because there is not enough chemical present to overcome the defense mechanisms of the cell, organ, or system. Above this threshold level, toxic effects may occur because the defense mechanisms are overwhelmed.

As an analogy, consider a dam along a river that starts in the mountains. The dam always holds back a certain amount of water, even in sunny weather. When the snow melts in the spring, the amount of water reaching the dam goes up. However, this water doesn't just pour over the dam and cause flooding to downstream areas. Instead, spillways and flow rates are adjusted and the amount of water that flows across the dam is controlled at a safe level. However, if a series of torrential storms occurs during warm weather in the early spring, after a very snowy winter, the amount of water reaching the dam may be too great for the spillways' control measures to handle. As a result, too much water may get over the dam, leading to flooding. In this analogy, flooding corresponds to the toxic effect. The spillways and control measures would be our internal defense mechanisms, and water would be the chemical. This simple example illustrates that a variety of events must occur before enough chemical reaches a target to cause toxic effects.

Species Variability

Different species react differently to chemicals. Chemicals that we consider safe may be toxic to other species. For example, dogs lack an enzyme needed to metabolize chocolate. Therefore, chocolate is toxic to dogs. Obviously it isn't to humans, or else the human race would be in big trouble! Even for chemicals that are toxic to most species, there are tremendous differences in sensitivity of toxic effects across species. A great number of chemicals are used as pesticides, fungicides, herbicides, etc. precisely because these chemicals are more toxic to a weed or insect species than they are to humans. The entire basis of the use of pesticides is differential sensitivity of species to chemicals.

As a common example, think of the spray you use to kill ants, roaches, or other household insects. We spray this on ants and they die within seconds. However, if you spray the solution on your hand, you merely wash it off and don't typically notice any effect. This illustrates the extreme range of chemical sensitivity across species.

This range of sensitivity across species is very important, particularly as it relates to identifying target or "safe" levels for humans by using laboratory animals. These laboratory animals are usually mammals, because we also are mammals. Therefore, our physiologies are more similar than if other species, such as frogs, were used. However, species sensitivities range widely within mammals as well. Historically we used rotenone as a fish poison because rotenone was highly toxic to fish but not very toxic to humans. Why is this so? There are diverse factors that govern species sensitivity to chemical toxicity. These include anatomy, physiology, metabolism, age, sex, health and nutritional status (see Chapter 6, Species Differences, for further discussion). For example, many laboratory mammals have been tested for toxicity from exposure to dioxins, including the mouse, rat, rabbit, dog, hamster, and guinea pig. The guinea pig is about 250 times more sensitive to dioxin than any other species tested, and 5,000 times more sensitive than the hamster. Using the most sensitive species in setting standards for protection of human health from chemical exposure is another way that toxicology concepts drive our regulations. This approach minimizes the chance that we may be more sensitive to a chemical than the other species on which the chemicals have been tested.

Dose-Response

This is typically the primary thrust of basic toxicology courses. The dose-response relationship defines the potency of a chemical. It relates the amount of chemical to a specific effect. As we discussed above, there are two types of basic dose-response relationships: threshold and non-threshold. These are illustrated in Figure 3.1. As shown in the figure, any dose can cause an effect in a non-threshold relationship (Chemical Y). As the dose increases, the relative effect will also increase. For a threshold relationship, which is the more common of the two across all chemicals, effects only occur above a certain dose (Chemical X). Above the threshold level, there is little difference between the two relationships. At high doses, the lines are essentially parallel, indicating that the response changes at the same rate for both chemicals. The rate at which effects increase with dose defines the potency of a chemical. This is the slope of the line in Figure 3.1. Using this principle, you can compare the potencies of different chemicals by comparing their slopes. It is easier to estimate the slope of a straight line than the complex lines shown in Figure 3.1. Therefore, these dose-response relationships are typically presented using a logarithmic scale for dose (the x-axis). This produces a straight-line relationship between dose and response, as shown in Figure 3.2. Using this figure, it is apparent that the slopes of the two lines are about the same. This indicates that both chemicals have the same potency. However, effects can occur with chemical Y at lower levels than with chemical X because they have different types of relationships (i.e., non-threshold and threshold, respectively).

If two chemicals have the same potency, one can still be considered more "toxic" than the other, as is evident in Figure 3.2. Even though the lines are parallel, a higher percentage of responses for chemical Y occurs at every dose level than for chemical X. For example, at a dose of 1000, there are no responses seen for chemical X; almost 50 percent responses are reported for chemical Y at this dose level.

Another factor that complicates how to interpret "potency" concerns dose-response curves with different slopes, such as the two chemicals shown in Figure 3.3. Such disparate shapes can result when the mechanism of action between the two chemicals is different. If asked to determine whether chemical X or Y is more toxic, how could you respond? It depends on the dose! At a dose of 2700 (B in Figure 3.3), the two chemicals have the same toxicity; the same dose leads to equal frequency of responses for both chemicals. However, at lower doses such as 700 (A in Figure 3.3), chemical Y is far more toxic—more than double the responses of chemical X for the same dose. The situation is reversed at high doses (C in Figure 3.3)—chemical X is more toxic than chemical Y. If asked which chemical is more potent, you could truthfully answer that you couldn't tell; however, you could say which is more toxic at a given dose level.

This complication makes it difficult to use terms such as "potent chemical" or "highly toxic" without providing a reference point. Potency is only meaningful if used in comparison with another chemical (for example—strychnine is more potent than table salt). Similarly, "highly toxic" is a subjective term when used alone. If used in comparison with other chemicals (dioxin is highly toxic to guinea pigs compared to hamsters), the term has more meaning.

The relationships discussed above apply to chemicals that exert toxicity at certain levels and have no therapeutic effects. There are other chemicals that are considered essential (e.g., Vitamin C, zinc) that may cause toxic effects if there is either too much (excess) or too little (deficiency) in the body. There is a window between these levels where the essential chemical has no toxicity.

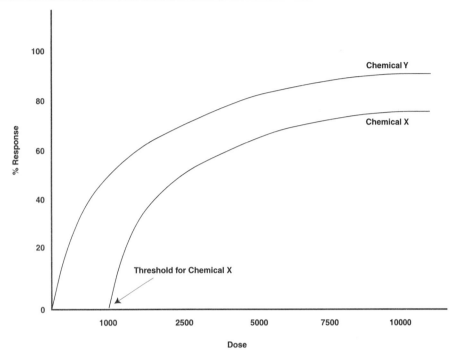

Figure 3.1. Sample Dose-Response Curves for Threshold and Non-Threshold Responses

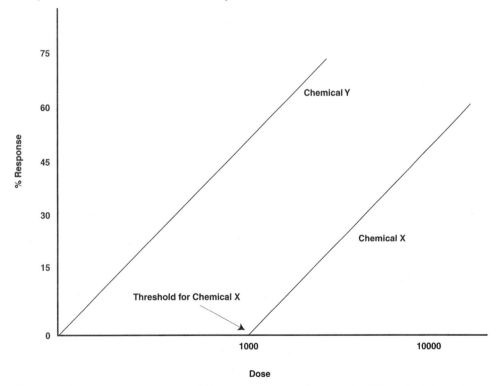

Figure 3.2. Sample Linearized Dose-Response Curves for Threshold and Non-Threshold Responses

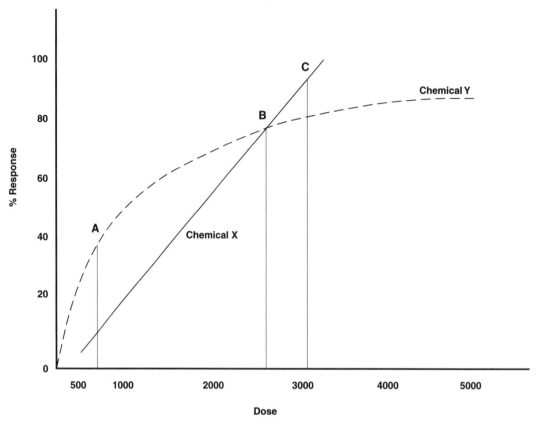

Figure 3.3. Which Chemical Is More Toxic?

With Vitamin C, which is ascorbic acid, we realized in the 1700s that deficiency could exert toxicity when English sailors began getting scurvy, which is characterized by bleeding from the gums, loosening teeth, and slow recovery from cuts and bruises. This disease was traced to a lack of fresh fruits on board ships during long voyages. This eliminated the main source of Vitamin C in their diet, and this led to a toxic deficiency. Vitamin C supplements are becoming part of our lives as more and more benefits are reported. Toxic effects have been reported from excessive levels, but only mild effects such as stomach upset from the acid.

Zinc is a component of several enzymes and other chemicals that are essential to our survival. Zinc deficiencies have been associated with many effects in the body, including effects on the skin (dermatitis, hair loss) and immune systems (lowered resistance to infections). Zinc supplements are common in stores, and we don't think about taking too much. However, at very high levels, zinc is a toxic metal that can affect the digestive tract (e.g., intestinal cramping, diarrhea). Many environmental cleanup sites must set "safe" levels of zinc that can remain in soil and won't lead to toxic effects from possible future exposure.

A sample dose-response relationship for an essential nutrient like zinc is shown in Figure 3.4. This illustrates an additional complexity to toxicology—effects can occur from too little as well as too much of a chemical.

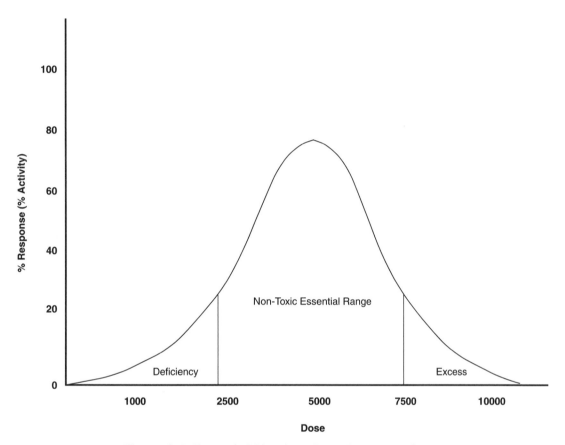

Figure 3.4. Essential Nutrient Dose-Response Curve

4

Types of Toxic Effects

This chapter builds on the basic concepts presented in Chapter 3 and discusses the types of cancer and non-cancer effects that can occur from chemical exposure. It provides an overview of specific effects of some well-known chemicals, and relates these effects to the mechanisms of toxic action discussed in Chapter 3. Effects on humans will be discussed first, followed by effects on other species (i.e., ecological effects).

Cancer

Cancer toxicity is exhibited by the formation of tumors. Cancer is defined as uncontrolled cell division. The vast majority of cancers are as yet unrelated to chemical exposure. Currently in the United States, approximately one in three (31 percent) people develop cancer over their lifetimes. Of those people developing cancer, the death rate is about 30 percent across all types of cancers. As of 1996, cancer was still the second leading cause of death in the United States (behind only heart disease). There are many types of cancers, which are typically classified both by the organ primarily affected (e.g., lung, breast) and the type of tumor produced. Although many types of cancers are often fatal, several types are typically not fatal (e.g., skin melanomas from ultraviolet light exposure due to sunlight).

Historically, it was thought that if a chemical could cause cancer, it could lead directly to tumor formation. However, it has more recently been learned that there are multiple stages to the development of cancer (see Chapter 3). As we learn more about specific chemicals, we are realizing that only a few chemicals are what are known as complete carcinogens; that is, one chemical can lead a tissue through all four stages of carcinogenesis. Many chemicals are known to be initiators (e.g., benzo(a)pyrene), and some are promoters (saccharin, phenol). For chemicals that are not complete carcinogens, other actions need to occur for cancer to develop from chemical exposure. For example, a chemical might initiate a cell to become cancerous. By itself, this does not cause cancer. Cells have internal repair mechanisms to fix such damage. However, if another agent, which could be a chemical, causes the initiated cell to be promoted (e.g., cell division occurs from the damaged cell), then tumor formation might occur.

This is one reason why cancer treatment is so difficult. There are many stages of carcinogenesis and many agents that act at different stages. In addition, there are multiple types of mechanisms at each stage, complicating drug therapy that often requires knowledge of such mechanisms to be effective.

Another differentiation between chemicals that might be considered carcinogenic lies in their general mode of action. Some chemicals directly act on DNA; these are called genotoxic chemicals. This interference with DNA can lead to changes in DNA by many different mechanisms, but all of these mechanisms eventually involve affects to DNA. Examples of such chemicals include nickel and chromium, which are elements of the earth's crust and are often used in industrial processes; vinyl chloride, an industrial chemical used to make polyvinyl chloride (PVC) and a breakdown product of common solvents such as trichloroethylene (TCE); and benzo(a)pyrene, a component of petroleum formed during burning of organic material.

Other chemicals do not directly act on DNA, but exert their carcinogenic effects by targeting other systems linked with protein production and use. These are known as epigenetic chemicals. Examples of these types of chemicals include asbestos, which causes a specific and rare form of lung cancer called mesothelioma; estradiol, the natural female hormone referred to as estrogen; saccharin, used as an artificial sweetener in the 1970s until it was reported in 1977 to lead to bladder tumors in mice; and ethanol, contained in all alcoholic beverages.

There is typically a long period of time between exposure and initiation of cancer to development of cancer to a point where it can be diagnosed. This period of time is referred to as a latency period. As an example, skin cancer following arsenic exposure typically takes about 15 years to develop. Once a cancer is diagnosed, it can be extremely difficult to trace the origin of the cancer to exposure to a specific chemical many years before. This is in part because many different chemicals may affect the same organ (see Table 4.1). Only when many people who share a similar task (e.g., making asbestos, working with benzene) contract similar cancers can a link between cause and effect be made.

We will look at a few examples of these types of effects that we can ascribe directly to chemicals to illustrate these different types of effects. Overall, there are very few chemicals or chemical mixtures that have been proven to cause cancer in humans. These chemicals and the primary affected organs are listed in Table 4.1.

Table 4.1. Chemicals Known to Cause Cancer in Humans[a]

Chemical Name	Sources/Uses	Main Tumor Site(s)
Arsenic	Natural element of rock/pesticides	lung, skin
Asbestos	Naturally occurs in some types of rock/ insulation materials	lung, gastrointestinal tract
Benzene	Solvent	blood, bone marrow
Benzidine	Synthesis of over 300 dyes in leather, paper, and textiles	bladder
Coke oven emissions	Iron/steel manufacturing	lung, kidney, prostate
Diethylstilbestrol (DES)	Fertility drug	breast
Direct black 38	Azo dye	liver
Direct blue	Azo dye	liver
Direct brown 95	Azo dye	liver
Hexavalent chromium	Natural element of rock/ pigments and dyes	lung
Nickel	Natural element of rock/plating industry	lung
Radon	Natural radioactive element	lung
Secondhand smoke	Tobacco products	lung
Vinyl chloride	Synthesis of polyvinyl chloride (PVC)	liver, lung

[a] Based on information from U.S. EPA (1998).

Lung Cancer

Lung cancer is the third leading cause of cancer in the U.S. (behind prostate and breast cancers). In spite of decades of focused research, the mortality rate for lung cancer is still higher than for other prevalent forms of cancer (Table 4.2). This mortality rate represents an improvement over 30 years ago, in part because we have begun to understand how chemical carcinogenesis works, which aids in early diagnosis and treatment. Although we have reduced our exposure to many carcinogenic chemicals through regulations and education, the incidence of some cancers, including lung cancer, continues to rise (over the last decade, this rise in lung cancer incidence has been mostly among females).

Table 4.2. Incidence and Mortality Rate for Some Forms of Cancer[a]

Type of Cancer	Incidence (Annual number of cancers per 1,000 people)	Mortality (Annual number of deaths per 1,000 people)	Ratio of Mortality to Incidence
All Cancers	4.41	1.29	0.29
Lung	0.69	0.38	0.55
Lymphoma (non Hodgkin's)	0.145	0.054	0.37
Colon/rectum	0.49	0.12	0.24
Skin Melanoma	0.116	0.019	0.16
Breast	1.07	0.113	0.11
Prostrate	1.85	0.062	0.03

[a] Based on 1996 data from the Centers for Disease Control and Prevention (1998), *Monthly Vital Statistics Report*, Vol. 46, No. 12(S)2, September, 1998.

Much of the information regarding how cancer develops and how it can be treated was painfully learned only after humans began getting the same types of cancer from exposure to a given chemical, or mixture of chemicals, over time. Examples of these chemicals are discussed below.

Asbestos manufacturing and use was big business in the years when the space program was on its course to the moon. Asbestos was a great insulator and was used for everything from wrapping pipes and ceiling tiles to hair dryers and pot holders. However, after many years of manufacturing, workers began to develop a specific, novel type of lung cancer called mesothelioma. The type of lesion had almost never been seen before, and was only present in those workers involved with asbestos handling.

For example, a 1976 epidemiologic study of over 19,000 asbestos insulation workers in the United States, Canada, and Ireland reported 182 deaths from mesothelioma in these workers; none were expected based on the historic incidence of this type of cancer worldwide. This led to other studies of workers in the textile and friction products manufacturing industries, asbestos cement product industry, and mining and milling of naturally occurring asbestos in rock, which showed similar increased incidence of this type of cancer. It was then determined that asbestos fibers were inhaled, lodged in the lung airways, and led to development of cancer at these locations within the lung. Based on this evidence, the U.S. EPA listed asbestos as a Group A chemical, which is defined as a known human carcinogen (Table 4.1).

Subsequently, asbestos has been removed from thousands of buildings across the country, including many schools. Asbestos only poses a health threat when the material is flaky or powdery, and small particles can float in the air and be inhaled. In its intact, natural form, it is inert and nontoxic. Although removing asbestos from buildings might make the buildings safer for its occupants (depending on the status of the material), it presents a risk for the workers removing the asbestos. The process of removal makes the asbestos flake, increasing exposure for these people. Therefore, a respirator is required for workers involved with asbestos remediation. Again, this is an example of a regulation designed to minimize exposure, and therefore toxicity, of a chemical to humans.

Cigarettes have been known for decades to cause lung cancer. Cigarettes contain over 4,000 ingredients. Over 40 of these ingredients are substances that are known to cause cancer in humans or animals; some of these substances are present in tobacco and some are present as additives used by tobacco companies (Table 4.3). However, long before we knew which components of cigarettes were responsible, there was substantial evidence across the world that cigarette smokers got lung cancer more often than non-smokers. Filters helped reduce the tar intake from smoking, but the small particulates and vapors, which contain some carcinogenic chemicals, still get absorbed into the lung.

Table 4.3. Subset of Known and Likely Carcinogenic Compounds in Cigarettes[a]

Chemical	Human Cancer Status U.S. EPA[b]	IARC[c]
2-Naphthylamine	Probable	Known
4-Aminobiphenyl	Probable	Known
Benz(a)anthracene	Probable	Probable
Benzene	Known	Known
Benzo(a)pyrene	Probable	Probable
Cadmium	Probable	Known
Catechol	Probable	Possible
Formaldehyde	Probable	Probable
Hydrazine	Probable	Possible
N-Nitrosodiethanolamine	Probable	Possible
N-Nitrosodimethylamine	Probable	Probable
N-Nitrosonornicotine	Probable	Possible
N-Nitrosopyrrolidine	Probable	Possible
Nickel	Known	Known
o-Toluidine	Probable	Probable
Polonium-210	Known	Known
Tar	Probable	Known

[a] Based on information from U.S. EPA, 1998, the National Institute of Occupational Safety and Health (NIOSH), 1998, and IARC, 1998.
[b] United States Environmental Protection Agency, 1999.
[c] International Agency for Research on Cancer, 1999.

Due to the difficulties involved with identifying chemical causes of cancer discussed earlier (e.g., latency time), a definite link between second-hand cigarette smoke, also referred to as environmental tobacco smoke (ETS), has only recently been established. Both the U.S. EPA and IARC have recently classified ETS as a known human carcinogen (Table 4.1). This should be expected because ETS is chemically similar to the smoke inhaled by smokers, which includes several carcinogenic chemicals.

It is difficult to establish a link between chemical exposure in the environment and cancer because of the latency period and the lack of specific information on the level of exposure (i.e., the dose) someone might receive second-hand. This is more difficult than evaluating risks to the smoker alone, because the amount of exposure is difficult to estimate. For example, you can't ask someone how many packs of cigarettes they inhale from the environment if they don't smoke. You can, however, identify people with lung cancer that have never smoked, and find out if they grew up with smokers in the house, or worked with smokers. Linking effects to possible causes using this approach is the basis of the science of epidemiology. To illustrate the use of epidemiological research in toxicology, the process used by the U.S. EPA to classify ETS as a known human carcinogen is summarized below.

U.S. EPA first looked at 30 available studies of the lung cancer incidence of non-smoking women whose husbands were smokers. In 24 of these studies, there was an increased risk of lung cancer in the women over those whose husbands did not smoke. Less than half of these studies reported a statistically significant increase. However, the probability that this many studies would find significantly higher risks is less than 1 in 10,000.

This first comparison includes those women that might have received only little exposure from their husbands. This is the problem of quantifying exposure to environmental chemicals discussed above. If these women did not get lung cancer, then this might artificially lower the actual increased risk to ETS. To account for this unknown, U.S. EPA looked only at the 17 of these 24 studies in which the men smoked the most. When incorporating the rate of exposure into the evaluation, all 17 studies showed an increase in lung cancer risk from ETS. Nine of these 17 studies reported significantly increased risk. The probability of this many studies finding significantly higher risks is less than 1 in ten million. This is less likely than winning the state lottery or being struck by lightning. At this point, the evidence becomes pretty strong that the increased risk of lung cancer is directly related to ETS. However, the U.S. EPA went one step further.

As discussed in Chapter 3, one key feature of toxicology is the dose-response relationship, which describes how increasing doses (e.g., exposure) result in increased incidence of response (e.g., cancer). The U.S. EPA looked for such a relationship between the level of ETS and lung cancer incidence in the wives whose husbands smoked. Fourteen of the 24 studies provided dose-response relationship information. All 14 of these studies showed increasing incidence of lung cancer with increasing levels of ETS. Ten of these 14 studies reported significantly increased risk. The probability of this happening by chance is less than 1 in one billion. This provides very strong evidence of causality.

As is evident from this example, the process starts with identifying general trends, getting more and more specific as you try to eliminate other sources or causes from consideration. The process usually doesn't work as clearly as shown above. As the U.S. EPA reports, it is unprecedented for such consistent results to be seen in epidemiological studies of cancer from environmental levels of a chemical or chemical mixture. In other words, it is rare that epidemiology alone can be used to link cause and effect. This is due to the many variables that are typically not consistent across all the people included in the studies. These variables that influence toxicity are discussed in Chapter 6.

If cigarettes were dumped into the environment and sprinkled on the ground and regulated as industrial chemicals, the cancer risk from possible exposure to the carcinogenic chemicals contained in cigarettes would require them to be cleaned up and disposed of in a landfill. However, because exposure to chemicals from smoking cigarettes is voluntary, any cancer risks from smoking are accepted by the individual. However, with regard to second-hand smoke, the person exposed is in an involuntary position. When exposures are not voluntary, people usually react by wanting the exposure to be eliminated. This difference between voluntary and involuntary risks lies at the heart of much of the confusion, fear, and emotionalism associated with risk assessment, toxicity, and exposure to chemicals.

Skin Cancer

The lung cancers discussed above represent a form of cancer with a high lethality rate. Some cancers, such as some forms of skin cancer, are not particularly lethal; the death rate from skin cancers is less than most other forms of cancer. Skin cancers are tumors of the skin, including those associated with abnormal growth of moles or discolorations of the skin.

Two examples of causes of skin cancer are exposure to ultraviolet light and ingestion. Skin melanomas often form in our middle years in response to many years of exposure to ultraviolet light while outdoors. These melanomas do not pose an appreciable risk of dying, and can usually be treated with topical ointments or creams. As shown on Table 4.2, about 3 percent of all cancers are skin melanomas. The risk of getting skin cancer is measured by the percentage of people that get skin cancer, the same way the risk of getting more lethal cancers (e.g., lung cancer) is measured. However, the risk of getting these two forms of cancer is not the same as the risk of dying from the cancer. As is shown on Table 4.2, the risk of dying from lung cancer (about 55 percent) is much higher than that for skin melanomas (about 16 percent). Chemicals causing both types of cancers are regulated identically. This makes it especially important to understand the basis of statements about risk to determine how to apply the information.

Ingestion of high enough levels of arsenic, usually from drinking water, can also lead to development of skin cancer. This is different from the skin cancer due to ultraviolet light because this type of skin cancer usually forms in areas not typically exposed to sunlight. Also, while ultraviolet light causes skin cancer from direct exposure, arsenic causes skin cancer only when ingested. Although this form of cancer is also usually not lethal, the amount of arsenic in our drinking water is regulated, so no toxicity should be associated with consumption.

This brings up an interesting observation. Regulations based on cancer chemicals do not differentiate between the types of cancer, for example lung cancer versus skin cancer. Even though lung cancer results in a much higher death rate, chemicals causing skin cancer are regulated to the same level of risk as all other carcinogens. It would be more useful if the regulatory values used for carcinogens reflected the severity of the cancer. In this way, chemicals causing cancers that have high survival rates could be regulated less stringently than those with lower rates of survival. This would allow for our regulations to focus most on those chemicals posing the greatest risk to our survival, while still protecting the public from all regulated chemicals.

Leukemia (Blood system)

Leukemia is cancer of the bone marrow that affects white blood cells; white blood cells help protect us from infection and disease. There are multiple types of leukemia, each differing in its lethality. The incidence of leukemia is about the same as that for skin melanomas. However, the death rate from leukemia (about 45 percent) is much higher than that for skin melanomas (about 16 percent).

One of the first chemicals definitively linked to a specific cancer in humans was benzene. Benzene was historically used as a solvent, and is a component of gasoline. After many years of exposure, workers in the buildings where benzene was used as a solvent began to develop leukemia. For some of these workers, leukemia didn't develop until years after they left the job. Using epidemiological evidence, researchers determined that the incidence of this type of leukemia was clustered around the workers at these plants that used benzene. With subsequent studies on animals, this link between benzene exposure and leukemia was confirmed. Benzene levels are now regulated in most products. The amount of benzene we are exposed to when filling our gas tanks is small; however, exposure over the course of a lifetime does increase our risk of leukemia. Whether or not this increased risk is meaningful depends on one's perception of "acceptable risk," which we will discuss in chapter 8.

Breast Cancer

The incidence of breast cancer has risen sharply over the last several years. It is currently the leading cause of cancer in women. As yet, we have not determined many of the causes of breast cancer, but chemical exposure cannot be ruled out. Many chemicals have recently been shown to have endocrine disruption effects; that is, effects that modify the actions of our hormones. Because breast tissue has receptors for some hormones, chemicals that affect these hormones can affect the breast. It is not yet known if exposure to chemicals in the environment increases our risk for breast cancer. Other explanations for the higher rate of breast cancer recently include that more cancer cases are reported as people live longer and that women are better at detecting the cancer at early stages and are reporting this to their doctors. Endocrine disruption represents a new branch of environmental toxicology, and is in its infancy in terms of our knowledge and understanding. We will discuss this topic in more detail in Chapter 10.

Prostate Cancer

The incidence of prostate cancer has risen sharply over the last several years. It is currently the leading cause of cancer in men, and is twice as prevalent in men as breast cancer is in women. However, as with breast cancer, we do not understand the specific causes of prostate cancer. Chemical exposure might increase the risk of prostate cancer, but as yet we have rarely linked exposure to any specific chemical or chemical mixture with prostate cancer. One exception is emissions from coke ovens (see Table 4.1). These emissions contain several PAHs. As previously discussed, several specific chemicals within this group are carcinogenic to the lung. However, there has been evidence in animals linking these emissions with prostate cancer. Other explanations for the increased incidence of this cancer type are the same as those discussed above for breast cancer.

Unknowns

The examples of breast and prostate cancer illustrate our current lack of knowledge related to chemical impacts on hormones (the endocrine system) and on mechanisms of specific cancers. We are only now beginning to understand some types of chemical effects on the immune system. One example is dioxin, which has been shown to compromise the immune function in many rodent species. This increases the chance for an individual to get ill, and decreases the speed and effectiveness of recovery. These are often subtle effects, and directly ascribing them to a specific chemical is very difficult. That is one reason why animals are used in toxicology research. We can control their environments well enough so that the only difference between different animals is the amount of a chemical they receive. Then it is easier to see small differences caused by the chemical. We will discuss this more in Chapter 5.

Other major organ systems targeted by chemicals (examples include chloroform and dioxins) include the liver and kidney. As we discussed earlier, these organs have special functions that increase their exposure to chemicals compared with most other organs. The metabolism that occurs in the liver can activate some chemicals, such as chloroform, to become toxic.

Non-Cancer Effects in Humans

As discussed in Chapter 3, non-cancer effects do not lead to uncontrolled cell division. Rather, they cause specific effects in organs. This is the basis for drug therapy; the neurotoxic effects (e.g., paralysis and death) discussed in Chapter 2 regarding tetrodotoxin are non-cancer effects. Non-cancer effects are typically divided into the following categories:

- Target organ effects
- Reproductive effects
- Developmental effects

Target Organ Effects

Liver

All organs can be affected by chemicals, but effects are most common in the liver, nervous system, and kidneys. An example of a liver toxin is acetaminophen (the active chemical in Tylenol). The liver is the first organ through which all ingested chemicals pass before they reach other organs. The liver is sometimes referred to as the hazardous waste site (or garbage dump) of the body because its prime roles are to recycle blood cells and metabolize chemicals to prevent them from exerting toxic effects in the body. As a result, some chemicals damage the liver if exposure is high enough (e.g., alcohol; acetaminophen). Sometimes the liver metabolizes chemicals that are not toxic in their environmental forms in an effort to increase their elimination from the body. The result of this metabolism can actually activate chemicals to exert toxicity (e.g., PAHs), or enhance their toxicity (e.g., chloroform). Then the metabolized chemicals go through the bloodstream and impact other organs (e.g., skin tumors from PAHs), or directly damage the organ in which it was metabolized (e.g., liver necrosis and liver cancer from chloroform). This phenomenon is known as metabolic activation. In the vast majority of cases, however, the liver decreases the toxicity of chemicals and increases their rate of elimination from the body.

For example, arsenic exposure can lead to lung cancer, skin cancer, or non-cancerous effects to the skin. Arsenic is a naturally occurring element, and historically was used in insecticides. Absorbed arsenic will reach the liver. Once in the liver, the arsenic is converted to a nontoxic form by an enzyme, known as methyltransferase, and is then excreted from the body.

In order for this enzyme to work, a chemical known as S-adenosylmethionine (SAM) must be present in the liver. SAM is a derivative of the amino acid methionine, which is essential to our health. There is only a certain amount of SAM in our liver that is available to combine with arsenic and render it nontoxic. If more arsenic enters the liver than the amount of available SAM can handle, then excess non-metabolized arsenic may accumulate in the liver and be distributed to other organs, causing toxic effects. This is known as metabolic saturation. As long as the amount of arsenic (i.e., the dose) is below this saturation level, our liver can detoxify it and no toxic effects will result. It is only when the amount of arsenic overwhelms the metabolic pathway that toxic effects can develop. This illustration of a threshold below which toxic effects do not occur is a key concept in toxicology (see Chapter 3 for a graphical illustration of the threshold concept).

Effects on the Nervous System

The nervous system is composed of two parts: peripheral and central. The peripheral nervous system controls our voluntary actions, such as moving our arms and legs and clenching a fist. Some chemicals affect only this portion of our nervous system, such as some organophosphorus insecticides. This system also controls our diaphragm, which in turn controls our breathing. Neurotoxins often impact the nerves of this muscle by mimicking natural chemicals used by our nerve cells (e.g., black widow and rattlesnake venom).

Chemicals like cocaine impact our central nervous system. The central nervous system controls our autonomic system and brain function. It is protected by a barrier that must be penetrated before it can be affected by chemicals. Many chemicals can pass through this blood-brain barrier (e.g., toluene from sniffing glue) and affect us. Many of the effects are anesthetic-like (e.g., chloroform), but others can impact learning (e.g., lead) or basic brain functions (e.g., tetanus toxin). Cocaine acts on specific receptors on nerve cells in our central nervous system by altering the level of neurotransmitter at the junctions between nerve cells. It is these neurotransmitters that allow for electrical signals to travel through our nervous system. Cocaine also inhibits the synthesis of adenosine triphosphate (ATP), a chemical used for energy that fuels our metabolism.

Effects on the Kidney

Because the kidney filters unwanted things out of our bloodstream and excretes them in urine, it is the target of unwanted chemicals. Many metals impact the kidney because they are particles that can block the filter, much as dust clogs air filters in your home. Examples of such metals include cadmium and mercury.

Reproductive Effects

Some chemicals impact the ability to reproduce. Chemicals may act on both sexes, or may only affect one sex. An example of a chemical that impacts reproductive ability of only one sex is dibromochloropropane (DBCP), a fumigant formerly used as a fungicide. Some male workers chronically exposed to relatively high levels of DBCP during synthesis and use became sterile. Sterility was not observed in female workers. It was learned that DBCP specifically affects Sertoli cells, which are located within the testis and protect, support, and nourish developing sperm. If enough Sertoli cells are damaged, sperm production is reduced, and if few enough sperm are produced, sterility can result. Other chemicals, including tri-o-cresyl phosphate (TOCP), an industrial chemical used as a plasticizer in lacquers and varnishes, directly affect the sperm (e.g., motility is reduced).

An example of a chemical that impacts the reproductive ability of females only is formaldehyde, an industrial chemical present in many building materials, including carpeting. Reports indicate that spontaneous abortions occur more frequently in mice administered high doses than mice given lower doses. We are not sure if this also occurs in humans. In addition to reproductive effects, this chemical can also cause stomach ulcerations, skin blemishes, and kidney damage.

An example of a chemical that impacts the reproductive ability of both sexes is ethanol, found in alcoholic beverages. In males all stages of sperm development within the testis are affected; in females ethanol appears to act by interfering with the synthesis of estrogen by the ovaries. In addition, we know that ethanol consumption during pregnancy can lead to deformities and fetal alcohol syndrome.

Developmental Effects

Some chemicals impact the developing fetus. These are known as developmental effects. They differ from reproductive effects because they do not impact the ability to conceive, but directly affect the fetus. There are two general types of developmental effects, those that lead to death of the fetus (known as fetotoxic effects) and those that impact the growth of the fetus (known as teratogenic effects). A teratogen is a chemical that causes a mutation in the DNA of a developing offspring. If the mutation is severe enough, death can result. If the mutation is less severe, the fetus will survive but will be adversely affected in some way. Developmental effects of chemicals are unique in that exposure is to the mother, but toxic effects occur to a fetus. Therefore, chemicals must pass through the placenta (in mammals) to reach the fetus, and then get to a target within that fetus. This historically made identifying such chemicals very difficult because effects were not seen in the adult animals upon which the chemicals were tested. This is painfully brought home in the following example.

A large increase in newborns with gross limb deformities was reported in West Germany in 1960. The children were born with either missing limbs or markedly reduced limbs due to lack of growth of the long bones of the arms and legs during development. In Hamburg, no cases of markedly reduced limbs in newborns were reported over the previous 20-year period. In 1961, the sedative thalidomide was linked as the causative chemical.

Thalidomide, introduced in 1956, was used throughout the world as a sleep aid and to decrease the nausea and vomiting associated with pregnancy. During animal testing prior to its release, it had no apparent toxic effects in humans or adult animals at levels used by the pregnant women. After the linkage between thalidomide and deformities was made, the chemical was withdrawn from the market in 1961, and no cases were reported after that time. Overall, about 5,850 malformed infants were born worldwide due to thalidomide use by pregnant women.

Based on this and other experiences, regulations within the United States and in many European countries now require that new chemicals, before they are released for use, be tested on pregnant mice or other laboratory mammals in order to evaluate their potential to cause toxic effects in developing offspring.

Ecological Effects

Chemicals also affect species other than humans. Most drug research is conducted on laboratory animals. Chemicals released into the environment can also affect wild species, such as DDT and eggshell thinning in raptors.

Toxic effects are not divided into cancer and non-cancer effects for wildlife species, primarily because cancer takes many years to develop and the shorter life span of most wildlife species, combined with predation, usually means that individuals do not die from cancer. Rather, effects are divided into different levels of ecological organization, which typically include the following:

- Organism-level effects
- Population-level effects
- Community-level effects

As you move from organism-level to higher level effects, you are in effect impacting a larger and larger group of organisms. This is analogous to going from the subcellular level to the organism level for a human. It is also analogous to human civilization in the following way. An organism-level effect can be considered equal to effects on an individual human. Population level can be analogous to effects on a neighborhood. Community level can be analogous to effects on a city. For ecological receptors, these different levels of organization involve multiple species in addition to just larger geographical areas (Figure 4.1). Let's look at ecological effects at these different levels using DDT as an example.

Organism-Level Effects

This is similar to how toxic effects are evaluated in humans. The target is the health of an individual organism. Using the DDT example, this would involve accumulating enough DDT to cause neurotoxic effects on a falcon. This requires fairly high levels; toxicity tests used to assess DDT's toxicity prior to its registration did not find that it caused a problem at the level of the individual.

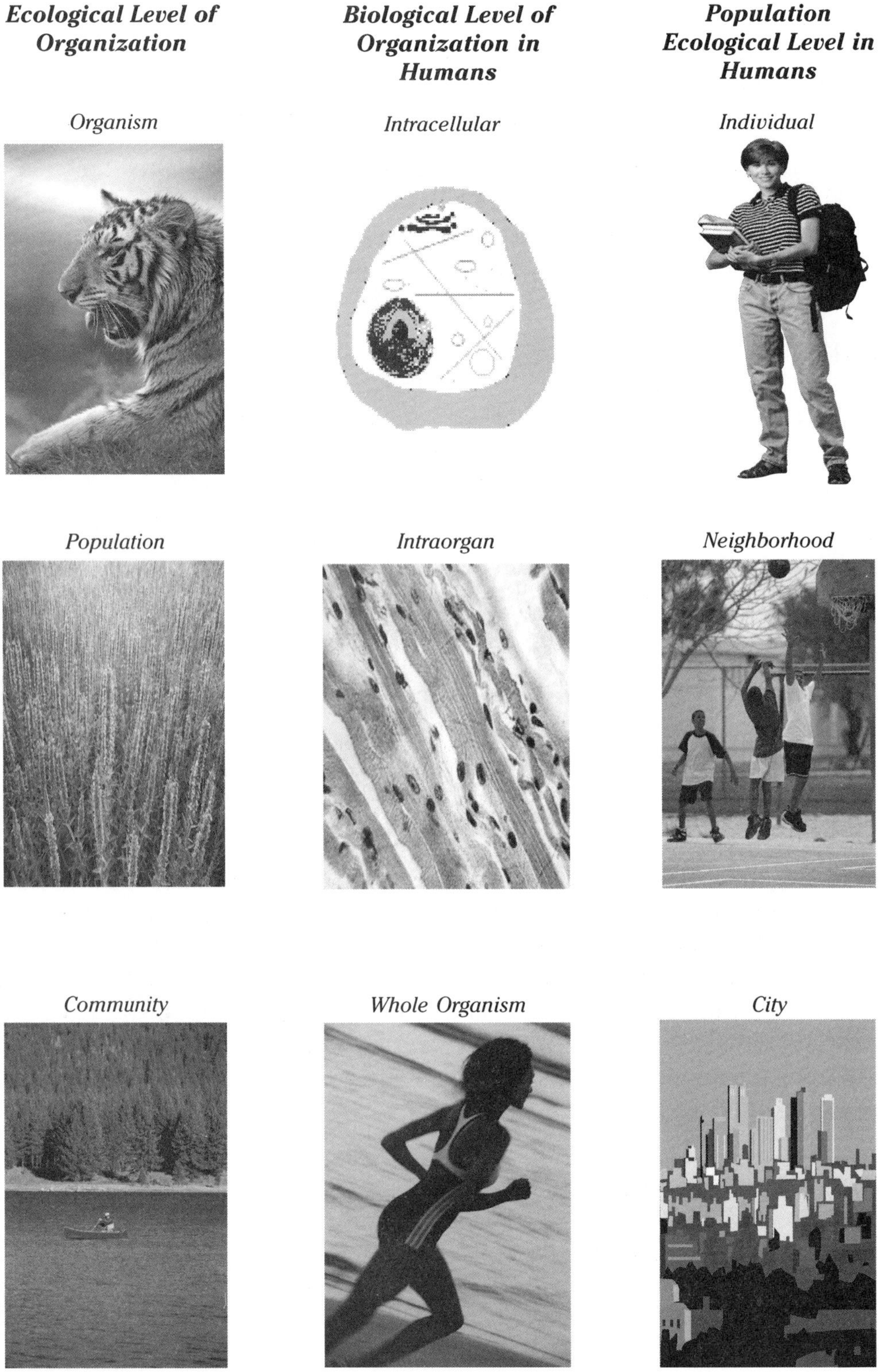

Ecological Level of Organization	*Biological Level of Organization in Humans*	*Population Ecological Level in Humans*
Organism	*Intracellular*	*Individual*
Population	*Intraorgan*	*Neighborhood*
Community	*Whole Organism*	*City*

Figure 4.1. Different Levels of Organization in Biology and Ecology

Population-Level Effects

This level focuses on effects on a population, which is defined as a group of individual organisms of one species in a contiguous geographical area. For example, this could be a prairie dog colony. Effects at this level of organization include effects on reproduction, because reproductive impairment can impact the number of animals in a population, which can then impair the health of the population as a whole. It is at this level of ecological organization that DDT exerts its effect on peregrine falcons. Eggshell thinning leads to decreased reproductive success, which impairs the population of falcons in an area impacted by DDT.

Community-Level Effects

This higher level of ecological organization incorporates all species in an area, and their interactions that maintain the health of the community. Using the DDT example, eggshell thinning at the population level may lead to fewer adult peregrine falcons. In turn, the lower number of falcons reduces predation on small mammals that supply the food base for falcons. As a result, the small mammal population explodes, leading to overgrazing of vegetation. This overgrazing impacts the food supply of the small mammals, which then cannot find enough food, leading to massive death of the small mammals. Even though the small mammals at both the individual and population level are not impacted by DDT, they are indirectly impacted based on a perturbation of the community structure.

This example illustrates the complexity of identifying toxic effects in nature. However, as with a human, there are built-in repair mechanisms that can overcome minor perturbations. Just as DNA can repair itself before cancer develops, effects to a few individuals, a population, or even a community can be corrected over a brief time period. It is only when the chemical exposure and toxicity overwhelm these repair mechanisms that permanent injury occurs.

For example, the DDT story ends nicely in that, some 25 years after the ban on DDT, the peregrine falcon has now been removed from the endangered species list. Since the banning of DDT, reproductive success has increased and the numbers of birds have recovered. If we had not banned DDT, these magnificent birds, unable to reproduce, might have become extinct. This illustrates the dynamic nature of our environment and its ability to heal itself if the cause of the problem is eliminated.

The higher the level of organization, the larger the injury needed to cause permanent damage. This is analogous to how we fight disease in our bodies. At the cellular level, many cells may die upon exposure to a virus (analogous to the organism level). This can lead to an organ infection (e.g., bladder infection; analogous to the population level). However, we can fight off this infection if it is not too severe, and there is no permanent impact to the organism (community level). If other factors compromise our ability to fight off the infection (e.g., AIDS, which reduces our immune system's ability to respond), then death can occur. Similarly, if other factors impact a community (e.g., other chemicals or habitat loss), the damage that could ordinarily be reversed becomes irreversible.

As you can see, the environment can act as an interactive "organism" encompassing many species, including humans. Battles similar to those fought internally in our bodies are fought on a larger scale by ecological communities.

Additional Reading

Klaassen, C. D., ed. 1996. *Casarett and Doull's Toxicology: The Basic Science of Poisons*. Fifth Edition. McGraw-Hill, New York.

5

Estimation of Toxic Effects

One important area of toxicology involves how toxic effects are actually measured. Humans don't typically get used as laboratory animals, yet the majority of our interests regarding chemical toxicity are related to effects on humans. Therefore, our understanding of human toxicology is largely based on studies of other species, such as laboratory animals. We have compiled some toxicology information from human exposures, which is referred to as epidemiology. Very rarely, we have used clinical humans studies (i.e., controlled studies similar to those used on laboratory animals) to get basic information on chemical effects. More recently, we have begun to use isolated tissues or cell types outside the body to study specific effects. These are referred to as *in vitro* studies, from the Latin phrase meaning "in glass." This chapter discusses the basic types of laboratory animal and cell culture studies used to quantify toxicity, and discusses how information on humans is used, when available, to quantify toxicity.

Regulatory Framework

Many of the toxicology tests discussed in this chapter are required under certain environmental laws, including the following United States Acts:

* Toxic Substances Control Act (TSCA)
* Federal Insecticide, Fungicide, and Rodenticide Act (FIFRA)
* Federal Food, Drug, and Cosmetic Act (FD&C Act)

Other Acts and regulations also require use of toxicology data, but the Acts listed above form the majority of required, defined animal testing methods and data for use across these other Acts. Each of these three Acts is briefly discussed below.

Toxic Substances Control Act

TSCA was enacted in 1977. Its primary purpose is to require that adequate data be compiled to determine toxic effects of chemicals posing "unreasonable risk." In effect, this includes about 50,000 chemicals currently in use, and about 1,000 new chemicals each year. Pesticides, food items or additives, drugs, cosmetics, tobacco, and radioactive chemicals are not included in these numbers because these are regulated under other Acts. For new chemicals, tests must conclude that the chemical poses "no unreasonable risk" based on animal testing. For existing chemicals, either a finding of "no unreasonable risk" is made, or certain restrictions or prohibitions on use are implemented for the chemical. For example, polychlorinated biphenyls (PCBs) were historically used in electrical transformers due to their physical properties. After TSCA was enacted, data indicating that PCBs could cause cancer in rodents at low levels led to prohibition of their use by 1981.

Federal Insecticide, Fungicide, and Rodenticide Act

This Act was originally enacted in 1947, but was revised in 1972 based on passage of the Environmental Pesticide Control Act. This Act covers all pesticides, not just the three types listed in the title of the Act. Under this Act, the manufacturer of a new pesticide must conduct a battery of toxicology tests to obtain registration for a specific use. If this specific use changes, additional testing must be conducted to evaluate toxicity of the chemical for this different use. For example, if a pesticide is approved only for use on cantelopes, more tests are required before it can be used on other crops like watermelons. Pesticides are classified into two categories by this Act: (1) those registered for general use (e.g., household), which are considered of relatively low hazard, and (2) those registered for restricted use (e.g., application on a specific crop by a certified pesticide applicator), which are considered of higher toxicity.

An overview of the regulatory requirements for animal testing of chemicals contained in these Acts is provided in Table 5.1.

Federal Food, Drug, and Cosmetic Act

This Act established the federal FDA, whose primary purpose is to protect the interest of the public by ensuring that foods, drugs, and cosmetics are safe, effective, and properly labeled. It established testing requirements for new drugs, which had to be proven safe for their intended use prior to marketing. It also required testing all food additives for toxicity. Some chemicals added to food are "generally recognized as safe" (GRAS). These chemicals are not considered food additives, and are not subject to the testing requirement of additives. In 1958, the Delaney Clause was added to this Act. The Delaney Clause states that any substance that causes cancer

Table 5.1. Regulatory Testing Requirements for New Pesticides in the United States[a]

Acute Tests	Subchronic Tests	Chronic Tests	Mutagenicity Tests	Bird and Mammal Tests	Aquatic Organism Tests
Oral Toxicity (LD_{50})	Oral Dosing	Feeding (Dietary)	Gene Mutation Assay (e.g., Ames Test)	Bird Single Dose (LD_{50})	Fish Acute (LC_{50})
Dermal Toxicity	21-Day Dermal Toxicity	Cancer		Bird Dietary (LC_{50})	Acute Toxicity to Invertebrates
Inhalation Toxicity	90-Day Dermal Toxicity	Teratogenicity	Heritable Chromosome Mutations	Mammalian Acute Toxicity	
Eye Irritation	Inhalation Toxicity	Reproduction		Bird Reproduction	Acute Toxicity to Marine Organisms
Skin Irritation	Neurotoxicity		Effects on DNA Repair	Field Testing for Mammals and Birds	
Skin Sensitization					Embryo Larvae and Life-Cycle Studies of Fish and Invertebrates
Delayed Neurotoxicity					Toxicity and Residue Studies

LD_{50} = Lethal dose resulting in death of 50% of tested animals.
LC_{50} = Lethal concentration resulting in death of 50% of tested animals.

[a] As required by the Federal Insecticide, Fungicide, and Rodenticide Act (FIFRA).

in animals or humans cannot be used in food in any quantity. This Clause was first used to prohibit the use of saccharin in food because the additive was shown to cause cancer in mice.

One reason why toxicity tests are required under these and other regulations is so that standardized testing methods and procedures (known as good lab practices, or GLP) will be followed for using chemicals in the marketplace. This ensures that the chemicals will be adequately tested, the methods used will be scientifically defensible, and the results will provide useful information regarding the toxicity of a chemical under its intended use to humans. This standardized protocol also prevents fraudulent experiments that are intended to demonstrate that a chemical is safe when it in fact is not. It is not possible to prove a chemical is safe, because this means proving that the chemical cannot cause toxicity. This is like trying to prove a negative. For example, you might be pretty sure that it has never rained in a certain part of a desert over the last 100 years because of direct measurements, but this is insufficient to say that it will never rain in that location. However, it is possible to show that a chemical does not cause toxic effects within the testing guidelines defined in the Act. This should indicate that the chemical is likely to be safe for its intended use.

The objective of this chapter is to familiarize the reader with the requirements for testing a chemical before it can be approved for use, and the types of tests used to meet those requirements. There are many types of toxicity tests that are not discussed in this chapter; the reader is referred to the Additional Reading section at the end of this chapter for more information regarding the field of toxicity testing.

The primary objective of toxicity testing is to minimize the potential harm to humans from chemical use. The specific objectives typically include:

- Identifying the target organ, target organs, or target system for the chemical
- Establishing if the effects are reversible or permanent
- Determining the most sensitive method for detecting the toxic effect
- Determining the mechanism of toxic action

Laboratory animal testing is still a common method for gathering toxicity information relative to these overall objectives. Although *in vitro* tests are becoming more prevalent in the scientific community, there is still uncertainty related to how tests on cells or tissues in a laboratory will translate to whole organisms exposed to the chemical in the environment. Recent information regarding reproductive effects that can occur at concentrations below those previously shown to cause toxicity based on laboratory studies (e.g., endocrine disruptors; see Chapter 10) also indicates that, for some information, animal testing provides the only reliable and accurate method for toxicity testing. The types of animal tests are generally discussed below, followed by a discussion of *in vitro* tests, human clinical trials, and epidemiology.

Laboratory Animal Studies

Laboratory animals are typically used to fulfill requirements of environmental regulations needed to register chemicals for use in the environment. Generally, animal tests can be categorized by their length. Using generally accepted protocols, three different lengths (or durations) have been developed:

- Acute tests (one-time dose or no more than one week)
- Subchronic tests (from one week up to 10 percent of the life span for the species)
- Chronic tests (at least 10 percent, typically 50 percent or more, of the life span for the species)

In addition to categorizing test methods by length, they can be categorized by the endpoint upon which the study is based. For example, some chronic tests are designed to identify if the chemical can cause cancer. Other tests (e.g., reproductive and developmental) might target specific exposure periods (e.g., pregnancy) to evaluate effects on offspring. All of these are discussed below.

Acute Tests

Acute tests typically do not extensively evaluate the toxicity associated with a chemical, but focus on identifying if the chemical can kill or directly affect a target species from short-term exposure. Acute tests do not attempt to identify how toxicity occurs, but only measure if it does occur. Acute tests are included in the overall toxicology testing battery required under some of the Acts discussed earlier (the Regulatory Framework section at the beginning of this chapter). Table 5.2 lists a sample of the required components of the acute testing protocols.

Table 5.2. Sample Acute Toxicity Tests and Commonly Used Species

Type of Test	Commonly Used Species
Oral Toxicity (LD_{50})	Rat, rabbit
Dermal Toxicity	Rat, rabbit
Inhalation Toxicity	Rat
Eye Irritation	Rat, rabbit
Skin Irritation	Rabbit, guinea pig
Skin Sensitization	Guinea pig
Delayed Neurotoxicity[a]	Chicken

LD_{50} = Lethal dose resulting in death of 50% of tested animals.

[a] Typically conducted on organophosphate insecticides.

The most common acute test is the median lethal dose (LD_{50}) test. The LD_{50} is defined as the dose of a compound that causes death in 50 percent of test animals over a specified time period (e.g., 1-day). The appropriate strain and species of animal must be selected based on the objective of the test. For example, to test pesticides for effectiveness, the target species should be very sensitive to the effects of the chemical (i.e., the pesticide should have the desired effects at low dose levels). In many cases involving pesticides, the appropriate test species could be an insect. Results of an LD_{50} study on insects would not be relevant for estimating the potency of a chemical in humans. To evaluate the potential effects of this pesticide on humans, a relevant mammalian species should be tested. Ideally, the chemical should act the same in the test animal as it would in humans. A sensitive strain of animals is typically selected so that effects on sensitive individuals could be evaluated. This type of testing is designed so that the potency of the chemical to humans is not likely to be underestimated.

This test can be used to rank the potency of chemicals, but only at the relatively high concentrations needed to cause death. This point was illustrated in Chapter 3 regarding assumed linearity of dose-response curves in defining potency. As we discussed in Chapter 3, two chemicals with different dose-response curves could differ in their relative toxicity at lower dose levels. In addition to reporting lethality, an LD_{50} study should also note the following:

- Changes in rate and depth of breathing
- Changes in locomotive functions
- Onset of convulsions
- Impaired reflexes
- Abnormal excretion

This additional information can help identify non-lethal toxic effects that increase the utility of the test.

The LD_{50} typically applies only to the oral route of administration (e.g., water or diet). For tests involving inhalation, LC_{50} values are identified. The LC_{50} is defined as the concentration of a compound in air that causes death in 50 percent of test animals over a specified time period. For aquatic species, LC_{50} values are also used; however, in this case the values represent surface water concentrations.

Acute tests are also used to evaluate the toxicity through the dermal route of exposure (e.g., skin). The dermal route of exposure is the most common way in which humans are exposed to chemicals (e.g., cosmetics). Therefore, acute dermal tests are an important portion of a toxicity testing battery. Acute dermal tests include those that evaluate toxicity or detect direct irritation and the allergic potential of the chemical. Those that evaluate toxicity do so by detecting whether adverse effects occur to the skin, and if the chemical is absorbed through the dermal barrier into the body. The chemical is applied to shaven skin dissolved in liquid (typically corn oil) and then reapplied at certain intervals to maintain exposure to the chemical over the entire test duration. This duration will differ depending on the goals and objectives of the study, and can range from weeks to years. These types of studies are often called "skin painting" tests. The chemical needs to be applied, or painted, onto an intact skin area (i.e., no abrasions). Rats, rabbits, and guinea pigs are the most commonly used animals for these studies.

Those that detect irritation use the same basic approach discussed above for toxicity tests, but typically involve only a 24-hour test period. The chemical is applied beneath a gauze pad, which keeps the chemical from being ingested through licking. At the end of the 24-hour period, the pad is removed and the area is cleaned. The endpoint of these tests is redness or swelling of the skin where the chemical was applied.

Another type of acute dermal test involves skin sensitization. This is most commonly used for cosmetics. In this test, a chemical is applied to the skin, enters the body, and leads to formation of antigens that react to the presence of the chemical.

This is similar to how pollen causes histamine release in allergy sufferers. These antigens are proliferated by white blood cells in lymph nodes, and are then released throughout the lymphatic system. This leads to hypersensitivity to the chemical upon further use. This is tested in the laboratory by giving an animal a "challenging" dose about 10 days after the sensitization test. This dose is applied to a different area of the skin than in the first test. The guinea pig is the animal of choice in these tests because the guinea pig has been shown to respond to chemical hypersensitivity similarly to humans. Rabbits, although widely used for dermal tests, typically are more sensitive to the irritant effects of chemicals than are humans. Therefore, the United States National Research Council recommends the guinea pig for dermal tests, especially those focusing on irritation.

The last type of acute test we will discuss is the (in)famous Draize test. This is an acute test that evaluates potential eye irritation, and was developed for the cosmetic industry to ensure that products used in eye make-up would not irritate the eyes. In this test, the chemical is placed directly on the cornea of the test animal using an eyedropper. The scientist then observes the signs and describes any lesions seen in the cornea, iris, or elsewhere in the eye and surrounding tissue. The test is considered cruel by many because the effects of the chemical often include pain and swelling of the eyes. However, this is one situation where *in vitro* tests would not provide similar information because ocular inflammation and other signs would not be reproduced in a beaker.

Subchronic Tests

Subchronic tests typically last from 14 to 90 days, and include a much more thorough evaluation of toxicity than for acute tests. Animals are usually sacrificed at the end of the test period to allow for the evaluation. The toxic endpoints examined include blood chemistry, urinanalysis, and histopathologic (e.g., microscopic) changes in many organs. Table 5.3 lists a sample of the required components of the subchronic testing protocols.

There are three main uses of subchronic tests. These are 1) to set dose levels for chronic tests, 2) as regulatory requirements for registering drugs, and 3) as regulatory requirements for using food additives. Regarding the first of these uses, one goal of chronic animal tests is to establish dose-response relationships, as discussed in Chapter 3. To establish a dose-response relationship, at least three different dose levels are required. These dose levels need to be developed so that the highest dose doesn't kill the animals. Subchronic studies are typically used to identify these dose levels for use in chronic studies.

Table 5.3. Sample Subchronic Toxicity Tests and Commonly Used Species

Type of Test	Commonly Used Species
Oral Dosing[a]	Mouse, rat, dog
21-Day Dermal Toxicity[b]	Rat, rabbit
90-Day Dermal Toxicity[c]	Rat, rabbit
Inhalation Toxicity[d]	Rat
Neurotoxicity[e]	Chicken

[a] Used to establish dose intervals for chronic testing.
[b] Short-term evaluation of dermal toxicity.
[c] Longer-term evaluation of dermal toxicity.
[d] Exposure typically 5 days/week.
[e] Tissue extracts biochemically evaluated for effects.

Many drugs are prescribed for short-term use consistent with the time periods of exposure used in subchronic tests. Therefore, to reproduce the expected pattern of use by consumers, subchronic studies are often used in drug testing. Subchronic tests are also required for food additives, and results must indicate that the chemical is safe for its intended use in order to be registered for use.

Chronic Tests

Chronic studies provide the most information regarding potential toxicity from chemical exposure. The endpoints of interest in chronic tests are typically microscopic changes within tissues or cells (e.g., histopathology). These changes can include such varied effects as tumor formation (e.g., cancer), elevated levels of liver enzymes in the blood (a sign of liver damage), interference with blood clotting (e.g., lead), and many others. Chronic tests require sophisticated and extensive planning in order to meet GLP requirements. These tests are typically expensive and require extensive testing facilities with "clean" rooms (e.g., germ-free), negative pressure rooms (to keep vapors within the room and prevent cross-contamination), and other special features. Table 5.4 lists a sample of the required components of the chronic testing protocols.

The main thrust behind these extensive and specific requirements is to enable someone several years from now to reconstruct the history of any specific animal in any test, including dosing regimen, laboratory notes, and lesions and other toxic effects. Given that several hundred animals are required in any one test, this represents a significant level of effort and need for good record keeping.

Table 5.4. Sample Chronic Toxicity Tests and Commonly Used Species

Type of Test	Commonly Used Species
Feeding (Dietary)[a]	Rat and dog
Cancer[b]	Mouse and rat
Teratogenicity[c]	Rat and rabbit
Reproduction[d]	Rat

mg/kg = milligrams of chemical per kilogram of body weight.
NOAEL = No-observed adverse effect level.
LOAEL = Lowest-observed adverse effect level.

[a] Lifetime studies used to identify NOAELs and LOAELs for noncancer effects.
[b] Typically 2-year studies.
[c] Chemical administered to pregnant females to evaluate birth defect potential.
[c] Multi-generational studies to evaluate possible effects on reproductive success.

General Guidelines

Typically, two species are required for a test (e.g., mouse and rat). Both sexes must be tested in equal numbers. Three dose levels are required, plus control animals (i.e., those not given the chemical). Fifty to 100 animals are required per group, per dose, and per sex. For each species tested, this would include a minimum of 400 animals, broken down as follows:

* 50 animals for each of three dose levels plus a control group, for a total of 200 animals of a given sex
* 8 animals for each of three dose levels plus a control group for clinical laboratory testing, for a total of 32 animals of a given sex
* 2 sexes, which doubles the above number to 464 animals

For the second species, the same calculations apply. This means that a chronic test conducted under the guidelines presented above (and which are used for cancer studies) would require a minimum of 928 animals.

The scientific approach to a chronic study is to eliminate all variables among the animals except one, which should be exposure to the chemical under study. Typically, all test animals of a given species are of one specific strain, which reduces genetic variability among the animals. All of the animals are about the same age and are maintained identically in the laboratory. They are fed the same diet and the same water, and breathe the same air. For some studies, the chemical will be administered through a tube directly into the stomach of each animal (i.e., gavage). This is done to ensure that the total measured amount of the chemical reaches the stomach of the animals.

While these environmental controls are necessary to identify minuscule subcellular differences between animals, they limit the value of the test when applied to the real world where young, old, sick, and pregnant receptors will also be exposed. This is an inherent limitation of laboratory studies, even those as extensive as discussed here. Extrapolating laboratory results to the environment, whether on whole animals or tissues, involves uncertainty. Uncertainty should be lower using tests from whole animals than tissues because the differences between the laboratory and the environment are smaller for tests on whole animals.

Although chronic tests can be conducted to evaluate only target-organ specific effects (e.g., kidney damage), they are generally conducted to evaluate either cancer, reproductive, or developmental effects. The testing methods and features specific to these are discussed in the following sections.

Cancer Studies

The goal of a cancer study is to produce tumors if a chemical is capable of doing so. To do this, the highest dose that doesn't cause death (i.e., maximum tolerated dose or MTD) is used as the highest exposure concentration. Because the incidence of most cancers is very low at typical environmental concentrations, a very high dose is used in the laboratory to increase the likelihood of detecting an increase in tumor formation compared with background levels.

Also, a dose-response relationship needs to be identified for potentially carcinogenic chemicals. This is needed to establish a quantitative relationship between dose and cancer risk. Because different levels of response are associated with different dose levels, a wide range of dose levels is needed for such tests. Tumor formation should also be seen in animals from at least one other dose level. These requirements make the selection of dose intervals critical to the success of the test, as discussed earlier under the Subchronic Studies section. Even though development of cancer might be without a threshold, below a certain dose no measurable increase in tumors would be seen because of the limited number of animals used at each dose level.

The oral exposure route is most commonly used for cancer studies (e.g., diet or water). Inhalation studies are less common because some variables are much more difficult to control than in oral studies. For example, you can quantify the amount of chemical exposure from water or food through direct measurement of the amount consumed. You cannot similarly measure the exact amount of air an animal has breathed.

The duration of the test is also critical, because animals are sacrificed at the end of the test for histopathological evaluation. Cancer typically involves a latency period between the time the exposure occurs and the time the cancer develops. This latency period can be 25 years or more in humans. This represents about 1/3 of the average human life span. Rats and mice are usually used in cancer studies because they have fairly short life spans (3 to 5 years). This means that a 2-year exposure represents a large part of their lifetime, and should be enough time to see tumors if they were going to develop. This is the standard test duration for carcinogenicity test. However, there are exceptions to this assumption. Diesel fuel was for many years not considered to be carcinogenic, even though it is known to contain chemicals that are considered to be carcinogenic (e.g., PAHs). This information was based on four cancer studies lasting no more than 2 years. When 30-month cancer tests were conducted, tumor formation occurred in all seven different tests. This indicates that there is a longer time lag (e.g., latency) between exposure to diesel exhaust and cancer development in test animals than for most other carcinogens. The reasons for this are still under study. Now diesel exhaust is considered a carcinogen by U.S. EPA and IARC.

The toxicological evaluation at the end of the test period includes separate visual evaluation of about 47 organs and tissues plus lesions, and cross-sections from several organs (e.g., liver, kidney, heart, bladder) for histopathological evaluation (Table 5.4).

Reproductive Studies

Typical toxicity tests do not allow for evaluation of reproductive toxicity, which can be defined as effects on the ability to reproduce, or on the survivability of embryos, fetuses, or infants. Therefore, specific tests were developed to specifically address this important area of potential toxicity. The testing guidelines include separate tests to evaluate the development of the fetus, fertility of the adults, and general reproductive performance (Table 5.4). Each test requires a year or more to complete, and the tests are generally quite expensive.

In a typical reproductive study, one or both parents are administered the chemical during reproductive behavior (males) or during pregnancy (females). The offspring are examined for effects; the healthy offspring are then mated when mature, and their offspring are similarly examined. Exposure is continuous across all of these age periods. These multi-generational studies provide additional data that evaluates changes in toxicity over long-term exposure. Such a situation occurs in the environment for smog. We are exposed to smog, and our offspring are also exposed. Their offspring will also be exposed. These laboratory studies attempt to identify what effects could occur in each of these generations, and lead to regulatory levels that protect against such effects.

Developmental Studies

Developmental studies are often associated with reproductive studies, because the testing methods can be similar but the endpoints are different. As opposed to a reproductive study, a developmental study solely evaluates the toxicity to the off-spring (i.e., developing fetus or newborn). The general goal of developmental tests is to establish "safe" concentrations to which humans are without risk of developmental toxicity. This is usually the highest dose level that does not cause a significant increase in developmental effects in offspring of laboratory animals.

Under the Food, Drug, and Cosmetic Act, reproductive and developmental tests are included together and are divided into Segments I, II, and III. Segment I tests focus on evaluating fertility of males and females using reproductive tests. Males are exposed 10 weeks before mating, while females are exposed two weeks before mating. The goal of the exposure timing is to cover a complete cycle of sperm development or estrus. Segment II tests evaluate birth defects from parental exposure using developmental tests. Pregnant females are exposed to the chemical from the time the embryo begins to develop in the uterus until all organs are formed in the fetus. This is the most sensitive time period for development, and the fetus is most vulnerable to teratogenic effects of chemicals during this time period. Segment III tests are designed primarily to evaluate survival of offspring after birth. For these tests, chemical exposure begins during the last trimester of pregnancy, and continues through the period of lactation. Therefore, the offspring is exposed both *in utero* and through mother's milk.

Based on the results of these tests, the FDA places drugs into one of four categories depending on the potential risk for developmental effects. These range from Category A (no evidence of developmental effects in humans) to Category D (a known number of birth defects). The highest dose that does not cause toxic effects in the offspring is used in the risk assessment process to identify a "safe" level of exposure (Chapter 8).

In Vitro **Methods**

In vitro tests are those conducted on portions of an organism. This is typically a cell culture from an animal or plant, or a tissue from an organism grown in the laboratory. Many *in vitro* tests are designed to evaluate the potential for a chemical to cause cancer. However, such tests are specifically designed to detect mutagens, not carcinogens. Mutagens are chemicals that cause a mutation to DNA. However, this is only one of several steps needed for cancer to develop (see Chapter 4). A popular type of mutagen test is the Ames assay. The majority of chemicals testing positive in the Ames assay have been shown to be potential carcinogens in whole animal tests. Therefore, this assay is a good screening tool that can identify chemicals that should be further tested for possible cancer activity.

One of the chemicals that tested positive in the Ames test is aflatoxin. Aflatoxin is a natural toxin produced by a fungus that grows on some food crops (e.g., peanuts) during storage in warm, moist locations. Aflatoxin had an extremely strong response in this test, which along with results in whole animal tests conclusively led to the identification of aflatoxin as a human carcinogen. Based on these tests, aflatoxin levels are monitored and must be at extremely low levels for the produce to be released to consumers.

Other *in vitro* tests are designed to identify reproductive or developmental effects of chemicals. These tests can include a wide variety of cells or tissues, depending on the endpoint of interest, ranging from house fly larvae to human embryonic cells. A sample of the types of cells and tissues used in such *in vitro* tests is provided on Table 5.5.

In vitro methods are becoming more powerful and reliable as the methods are refined and improved. Fewer tests are conducted using whole animals today than there were 10 years ago.

Table 5.5. Sample *in vitro* Toxicity Tests[a]

Type of Test	Commonly Used Species
Ames Test for Mutagenicity [b]	bacteria (*Salmonella*) with mammal liver enzymes
Heritable Chromosome Damage[c]	housefly (*Drosophila* sp.), mouse cells
Repairable DNA Damage[d]	bacteria (*Escherichia coli*), rat liver cells, human cells

[a] Over 35 different assays are routinely used, including tests on bacteria, fungi, plant, housefly, mammal, and human cells.
[b] Primarily used to identify potential carcinogenic chemicals (90% of chemicals positive for mutagenicity via this test are also carcinogenic).
[c] Referred to as "translocation" tests; effects seen in germ (i.e., sex) cells.
[d] Repair-deficient bacterial strains used to maximize test sensitivity.

Human Clinical Studies

For new drugs, human clinical trials are conducted following the battery of animal and *in vitro* tests discussed earlier. Based on the results of the animal tests, some drugs show a high degree of efficacy and a low incidence of side effects. Human clinical trials will be conducted to evaluate if use of such drugs appears safe in humans. These trials are intended to demonstrate that a new drug will be effective for its intended use, and that side effects are uncommon and mild. Such studies are not part of environmental chemical toxicology testing batteries, mostly because they are not required under federal or state laws.

Epidemiology

Epidemiology is defined as the study of the incidence and prevalence of disease in large populations. The source and cause of the disease is also identified. The animal studies discussed earlier in this chapter are controlled experiments. Instead of using controlled experiments, epidemiology is based on findings from observations in the real world. Because it is not considered ethical to use humans as experimental animals, data from controlled experiments are not available.

Estimating the toxicity of chemicals in humans can be done using two basic approaches: (1) observations are made on humans directly through epidemiology, or (2) data from controlled animal studies are extrapolated to humans (see Chapter 7). The former of these approaches is the most desirable because extrapolation, which involves uncertainty, is not necessary.

Due to all of the confounding factors among different individuals, it is very difficult to conduct and interpret results of epidemiology studies. Unlike laboratory animals, humans all lead different lives and have a wide variety of genetics, medical histories, and other factors. This makes it very difficult to identify an exposed group of individuals and a control group for comparison and to minimize all other differences between the groups. This approach is even more difficult for identifying chemical-based sources of effects because the incidence rate for toxic effects is typically very low. This low incidence rate means that the background incidence rate (i.e., the frequency at which an adverse effect occurs without influence by chemicals) might be high enough to mask toxic effects unless a very large population is affected. For example, assume the incidence rate of toxicity for exposure to 1 part per million of a chemical in drinking water is 1 in 10,000 based on laboratory animal studies. If this chemical contaminated a drinking water supply that served 500 people, this incidence rate corresponds to $1/100^{th}$ of one case would occur. In other words, it is quite possible that no adverse effects would be seen because the population was too small. This does not mean that the chemical is not toxic at this concentration, but that the effects cannot be differentiated from the background. However, if the chemical contaminated a reservoir that served a city of 100,000, then several cases should appear if the incidence rate in humans is similar to that seen in animals. Even in this situation, only 10 cases might be expected. The background incidence might be 1 in 20,000, so 5 cases would be expected. Proving a link between chemical exposure and the additional 5 cases of this effect seen in this example would be extremely difficult due to the variability in the human population. Additionally, factors such as the latency period between the exposure and the development of cancer, the high background rate of some cancers (e.g., lung), and the inexact knowledge of exposure by individuals (see Chapter 3), further complicate the process.

Epidemiological studies targeted at quantitatively linking chemicals and specific toxic effects typically are either (1) episodic or (2) retrospective. Episodic studies are those that evaluate isolated cases of toxicity (typically cancer) and attempt to link these effects with specific environmental factors (e.g., a chemical).

Episodic studies include those that involve cancer "clusters." One such cluster involving the chlorinated solvent TCE was popularized in a book (*A Civil Action*) and a movie of the same name. Twelve cases of leukemia developed in children living within a one-half mile radius in a small town in Massachusetts. The incidence rate was more than seven times higher than the national average. TCE was identified in the groundwater used for the drinking supply in that area of the town. Therefore, epidemiological evidence suggested a link between the TCE and the leukemias. However, such a link was not proven in court. In part, this was due to the very small sample size for the study, which was only a small portion of the population's 35,000 people.

Retrospective studies review the history and habits of targeted populations with specific effects in an effort to identify a cause of the effects. These are the most typical types of epidemiological studies conducted in toxicology. Such studies have led to identifying the links between leukemia and benzene, lung cancer and smoking, mesothelioma and asbestos, etc. In these studies, two groups are typically compared. One group has specific toxic effects and exposure to the substance under evaluation. The other group has no toxic effects and was not exposed to the substance. All other differences between the two groups are minimized as much as possible. For example, both groups should be similar in regard to sex, age, medical history, and smoking habits.

Additional Reading

Klaassen, C. D., ed. 1996. *Casarett and Doull's Toxicology: The Basic Science of Poisons*. Fifth Edition. McGraw-Hill, New York.

Harr, J., 1995. *A Civil Action*. Vintage Books, A Division of Random House, Inc., New York.

Hayes, A. W. , 1984. *Principles and Methods of Toxicology*. Raven Press, New York.

6

Important Toxicological Parameters

So far, we have discussed the basic concept of a dose-response relationship, how toxic effects are measured, and the types of toxic effects that can occur in a variety of organisms. This chapter introduces many of the core concepts of toxicology. We have seen that toxicity can differ based on differences among species. The factors that result in these differences are further discussed here. The majority of these factors affect toxicity across species in predictable ways. Children and the elderly are often considered most susceptible to toxic effects of chemicals; the concepts presented here illustrate why this is so.

One key word defined in this chapter is toxicokinetics. Toxicokinetics can be defined as how a chemical acts in the body. This concept is discussed first, followed by a discussion of factors that affect toxicity within and across species. Since we saw in the last chapter that the majority of toxic effects are measured in species other than humans, and sometimes at the level of the individual cell, these concepts are key to understanding how toxic effects can be extrapolated to the typical human. Finally, this chapter discusses how a toxic effect is selected for use in quantifying the toxicity of a chemical to humans.

Toxicokinetics

Toxicokinetics is a subfield of toxicology and consists of the following four components:

- Absorption
- Distribution
- Metabolism
- Excretion

Toxicokinetics is also described as the *disposition* of a chemical. Absorption is defined as passage of a chemical across a membrane into the body. Until a chemical is absorbed, toxic effects are only rarely observed, and then only at points of contact with the body (e.g., acid burns on the skin). Once a chemical is absorbed into the body, it is distributed to certain organs via the circulatory system. How a chemical is distributed in the body in part governs the target organ for that chemical, how easily it is eliminated from the body, how long it may take to act, and how long and where it will persist in the body. Metabolism is the transformation of a chemical in the body to other chemicals. This can either increase or decrease the toxicity, but typically increases the water solubility of a chemical, which leads to increased excretion. Excretion is defined as elimination from the body, either as urine, feces, or through sweat or tears.

These four factors in combination govern the degree of toxicity, if any, from chemical exposure. For example, a chemical may be absorbed but never reach a concentration at the target location above a threshold level. Or the chemical might be absorbed at concentrations high enough to cause toxicity, but the chemical is not distributed to the organ in which it can cause toxicity. As a result, no toxicity will result even though the amount absorbed was above a threshold level. Alternatively, a chemical might reach a target organ in sufficiently high levels to cause toxicity, but specific activities of cells in that organ change the structure of the chemical so that it is less toxic. Again in this situation, a dose that is high enough to cause toxicity leads to no effects. Finally, a chemical could be eliminated from the body so fast that it never builds up in tissues enough to cause toxicity. These processes can occur simultaneously, and may either increase, decrease, or not affect the toxicity of a given chemical, depending on the magnitude and direction of each component. Each of these four toxicokinetic components is further discussed below.

Absorption

A chemical must be dissolved to be absorbed. Lipid (i.e., fat) solubility and size of the molecule govern the rate and overall degree of absorption for a given chemical. To understand how chemicals can be absorbed, an understanding of the anatomy of a cell is needed. Membranes within cells are semipermeable; chemicals can enter a cell through the membrane, but they need to diffuse through a sandwich of three layers composed of lipid, water, and lipid again to enter. Once in the cell, the chemical must then pass out of the cell on the other side. This may lead directly to the bloodstream, or other locations depending on where the chemical is absorbed. There are three main locations where chemicals can be absorbed. These include the lung, the intestinal tract, and the skin. The first two of these are designed to absorb oxygen and nutrients, respectively, while the third is designed as a barrier to absorption.

Lung Absorption

For absorption from the lungs to occur, a chemical must pass from the alveoli into the bloodstream. The alveoli are extremely thin structures; each is just one cell thick. Similarly, the blood vessels that line the alveoli are capillaries that are themselves only one cell thick. This design maximizes oxygen exchange between the lungs and the blood. However, this design also enables chemicals dissolved in air to pass into the blood as does oxygen. Smaller, lighter weight chemicals will be absorbed more efficiently because they more closely resemble oxygen than do larger chemicals.

The lining of our alveoli is moist; oxygen exchange could not otherwise occur. Therefore, some areas on the alveoli produce a surfactant, which is a fatty substance that lines the alveolar surface and keeps the thin structures from collapsing. Because this surfactant is primarily composed of fatty chemicals, fat-soluble (or lipid-soluble) chemicals will pass through the layer and into the cells more readily than water-soluble ones.

Before absorption can occur in the alveoli, the chemical needs to pass a host of defense mechanisms in the upper respiratory tract. These primarily include tiny hairs called cilia, which constantly beat upwards towards the mouth, and mucus, which is sticky and is secreted by cells in the bronchioles called goblet cells. Particles stick to the mucus, and the beating cilia act as an escalator and bring the particles up into the mouth where they are swallowed. These defenses are less effective if the lining of the respiratory tract has been damaged, such as by smoking. The damage kills cells lining the airways and destroys the cilia and mucus-producing glands. Therefore, inhalation absorption increases in situations where lung function has been reduced.

Gastrointestinal Absorption

Absorption through the gastrointestinal tract (GIT) includes absorption through the mouth, stomach, and intestines. Chemicals may reach the GIT from either direct ingestion or inhalation of particles and transfer of these particles from the respiratory tract to the mouth, as discussed above. The vast majority of absorption from the GIT is from the intestines, particularly the area of the small intestine known as the duodenum. The wall of the duodenum is highly folded into finger-like projections called villi. These villi are bathed in fluids from the GIT that assist in digestion and absorption. The villi are lined with a layer of epithelial cells that separates the GIT and the blood. The membranes of these cells contain digestive enzymes that help move chemicals into and through cells. These chemicals primarily act to increase the absorption of certain essential chemicals, such as amino acids. These mechanisms do not differentiate between nutrients and other chemicals, so potentially toxic chemicals could also be absorbed.

Once material reaches the bloodstream, the first destination is the liver. The liver serves as a filter for materials absorbed through the GIT. Toxic chemicals can be altered to nontoxic forms, or vice versa. This is referred to as the "first-pass effect"; chemicals can be altered by the liver before they reach their targets. This anatomical feature allows ingested materials to be screened prior to their release into the rest of the body.

Skin Absorption

Unlike the lungs and the intestines, the skin is not designed for absorption. The outer several layers of skin consist of dead tissue. The first living tissue encountered in the skin lies beneath these dead layers of skin, referred to as the stratum corneum. Therefore, a chemical must pass through several layers of cells to reach a blood vessel and enter the body. Also, blood vessels are spaced far apart in the skin, so even after a chemical reaches the living cell layers, it still might never reach a blood vessel to be transported to a site where it can cause toxic effects. These different types of absorption mechanisms are illustrated in Figure 6.1.

Overall, absorption is usually most efficient through the lungs and least efficient through the skin. Only 1 to 15 percent of a chemical amount is typically absorbed through the skin, depending on the type of chemical. However, more than 75 percent of an exposed amount is often absorbed from the lungs. The variability of absorption is highest for the GIT; absorption ranges from as low as 0.1 percent (for some metals such as cadmium) to as high as 100 percent (for some essential nutrients like calcium).

Distribution

Distribution defines where a chemical goes in the body once it has been absorbed. For some chemicals, distribution differs depending on how or where it was absorbed (e.g., skin or lung). For other chemicals, distribution is the same regardless of how and where absorption occurred. For example, regardless of how it was absorbed, the majority of lead that is absorbed will end up in our bones. However for cadmium, absorption by the lungs will lead to primary distribution in the lung, while absorption through the GIT will lead to distribution in the kidneys.

In addition to defining where a chemical goes, distribution also tells you what percentage of a chemical dose goes to different locations. For example, we know that lead can cause toxic effects in the nervous system and in the blood cells. However, the majority of absorbed lead (about 90 percent) eventually goes to the bone tissue. Not all of the lead goes to the bone, however. If it did, no lead would reach the targets of toxicity, and lead would not affect us. Therefore, the amount of a chemical that goes to a specific location is more important than what percent of the dose goes there.

Oral Absorption Across the Intestine

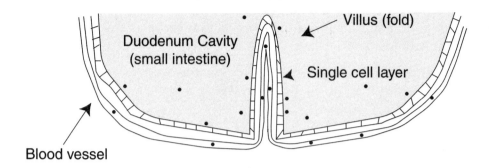

Inhalation Absorption Across the Lungs

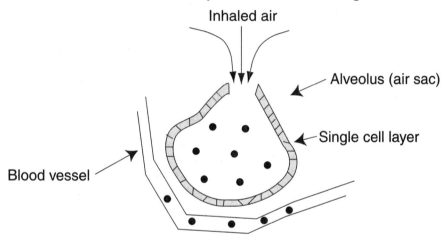

Dermal Absorption Across the Skin

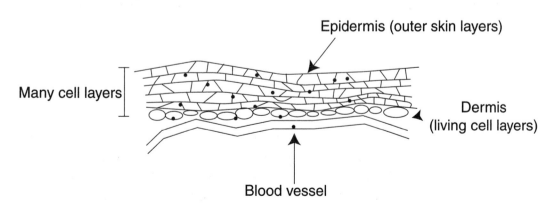

Figure 6.1. Absorption Mechanisms by Exposure Route

The primary factors that govern where a chemical will be distributed in the body include the blood flow to a particular tissue or organ, and the relative affinity of that tissue or organ for the chemical. Based on different combinations of these factors, there are four major locations in the body that serve as storage depots for chemicals. These include:

- Liver and kidney
- Fat
- Bone
- Plasma proteins

Each of these storage depots is further discussed to illustrate the relative importance of these two factors in determining where chemicals will go in the body.

Liver and Kidney Storage Depot

The liver and kidneys receive among the highest blood flows of any organs. For example, about twice as much blood flows to the liver than to the brain (see Table 6.1). The differences in blood flow are even more striking when the weight of the tissue is factored into the rate of blood flow. Table 6.1 presents a comparison of the blood flow to various organs and tissues as though they were all the same size. In Table 6.1 we see that the kidneys receive the highest relative blood flow other than the lungs, and about eight times as much as the brain. The liver receives the next highest blood flow. This means that chemicals in our bloodstream will more likely reach the liver and kidneys than other tissues or organs because more blood passes through these organs than most other organs. The liver is the primary site of chemical metabolism (see the Metabolism section below), so it is not surprising that blood flow is high to this organ. Similarly, the kidneys serve to filter unwanted items from our blood. To do this efficiently, the blood flow through the kidneys must be high in order to filter a lot of blood very quickly. As a result of this high blood flow and the types of activities that occur in these organs, the liver and kidneys are primary sites of chemical storage.

Fat Storage Depot

Fat has low blood flow. For example, 80 times more blood flows to the skin than to fat reserves (see Table 6.1). However, many chemicals are stored primarily in fatty tissues because they are lipophilic (fat-loving). For example, the distribution of DDT administered to laboratory mice through the diet was examined by chemically analyzing the organs. The results of this study indicated that roughly similar DDT concentrations were found in the brain and fat tissues, even though the brain receives 200 times higher relative blood flow than fat.

Table 6.1. Blood Flow to Various Organs and Tissues

Organ	Total Blood Flow (ml/min)	Organ Mass (% of body weight)[a]	Percent of Heart Output[b]	Relative Blood Flow (ml/min/100g)[c]	Percent Lipid	Relative DDT Concentration (mg/kg)[d]
Lungs	6000	1.4	100	600	3	2164
Kidneys	1200	0.4	23	400	3	87
Liver	1550	2.4	25	91	4	207
Intestines	1050	2.9	20	52	NA	not measured
Brain	750	2.1	15	50	8	40
Skin	400	2.9	9	20	NA	not measured
Fat	25	14	2	0.25	91	39

ml/min = milliliters of blood per minute.
ml/min/100g = milliliters of blood per minute per 100 grams of tissue.
mg/kg = milligrams of chemical per kilogram of body weight.
NA = Not applicable.

[a] Based on a 70 kg body weight for an adult human.
[b] Percent of blood from each heartbeat passing through the organ.
[c] Weight-normalized blood flow (assumes all organs are the same size [100g]).
[d] Measured in rabbits following DDT oral exposure.

However, because blood flow to the fat is very low, distribution into fat takes time. Therefore, a chemical like dioxin may first be distributed to the liver, but the amount in the liver will decrease with time as some reenters the blood supply and reaches fat cells. Some will also be metabolized by and interact with the liver, but the dioxin not affected by the liver will likely reach the fat cells. This phenomenon of a chemical having two different destinations over time is known as redistribution.

Blood flow and fat content are inversely proportional. In other words, as the blood flow to an area increases, the fat content of the area typically goes down. So, we have a situation where most chemicals want to be stored in fatty tissue, but the majority of blood is going elsewhere. Identifying the ultimate destination of chemicals in these circumstances is part of the discipline of toxicokinetics. The ultimate destination will likely change over time; the longer a lipophilic chemical exists in the body, the more likely it is to be present in fatty tissue. Once a chemical reaches fatty tissue, it can still reenter the general circulation if the fatty tissue is affected. Going back to the DDT example, it has been shown that short-term starvation has led to destruction of fatty tissues that liberated a high enough level of DDT to cause toxic effects. This demonstrates that fat may be a storage location, but this does not mean the chemical poses no threat in the body just because it is in the fat.

Bone Storage Depot

In addition to fat, bone is another tissue that has a relatively low blood flow yet still is the preferential storage location for some chemicals, most notably lead and fluoride. In general, chemicals that are stored in bone are elements (e.g., lead and strontium) that bind to the structure of the bone. As with fatty tissue, chemicals bound to bone can be redistributed if bone tissue is lost, such as in osteoporosis. Lead is not toxic to bone; however, both fluoride and strontium are.

Plasma Protein Storage Depot

Lastly, there are circulating proteins (e.g., albumin) in our bodies that function to transport other chemicals within our bloodstream. These plasma proteins are too large to be filtered through the kidneys or diffused out of the blood, so they remain circulating in the bloodstream. In essence, these circulating plasma proteins serve as a taxi service for chemicals. Some of the chemicals that use the service include Vitamin C, histamine, thyroxin, iron, and hemoglobin. These chemicals get to where they are going, and unbind from the proteins (i.e., get out of the taxi). These are the paying customers. Other chemicals go along for the ride, but are not recognized by the plasma protein. Some of these chemicals include drugs such as a sulfonamide or acetaminophen. These chemicals can take the place of paying customers and cause toxic reactions by either directly damaging the body or preventing the correct chemical from reaching its destination. For example, in the case of the sulfonamide, a treatment with a tetracycline and sulfonamide mixture was being used to deal with bacterial infections in premature infants. The mixture led to higher mortality of the infants than use of just the tetracycline alone because the sulfonamide displaced the liver byproduct bilirubin from plasma proteins. The bilirubin, unattached from the large protein, passed through the unformed blood-brain barrier of the infant and caused severe brain damage known as kernicterus. In this situation, the endogenous chemical, bilirubin, exerted the toxicity. However, the toxic agent was the sulfonamide because the action of the sulfonamide triggered the toxic response.

Barriers to Distribution

As referred to above, there are some barriers that affect where a chemical will be distributed. These include the blood-brain barrier and the placental and mammary (breast milk) barriers (in pregnant and nursing women). In mammals, the central nervous system (CNS), which includes the brain and spinal cord, is separated from the rest of the body by a barrier. This barrier is represented by a layer of special cells that lie between blood vessels and the CNS. These special cells do not allow chemicals or fluid to move between them; movement can only occur through the cells themselves. This greatly slows movement of chemicals from the bloodstream to the CNS, and allows time for intracellular actions to modify or destroy the chemical before it ever enters the CNS. However, many chemicals, especially lightweight gaseous chemicals like chloroform, can easily pass through the barrier because they dissolve in cellular lipids very well. Even chemicals that do not dissolve in fats, like lead, can pass through the barrier, but in a different way than more fat-soluble chemicals. These chemicals need to be modified to make them more lipid-soluble before they can pass through the barrier. For lead, it is methylated in the liver as part of regular metabolic processes. Addition of these methyl groups makes lead act much more like a natural chemical in our body and increases its ability to gain access to the CNS.

The CNS is bathed in a fluid known as the cerebrospinal fluid; this fluid is continuous throughout the CNS. This means that, once a chemical reaches the CNS, it will be distributed throughout the brain and spinal cord. Significantly, the blood-brain barrier is not developed at birth. Therefore, extremely young infants are particularly susceptible to chemicals because there is no barrier to prevent the chemicals from reaching the CNS.

The placenta exists in pregnant women, and separates the maternal and fetal blood supplies. Its main purpose is to pass nutrients and oxygen from the mother to the fetus, and to pass wastes from the fetus to the mother. This is similar to the actions of the lungs and the GIT; therefore, exchange of chemicals across the placenta is designed to be efficient. Unlike the blood-brain barrier, most chemicals can pass across the placenta either by simple diffusion or by mimicking necessary nutrients and using their active transport mechanisms. Cells within the placenta do have the ability to transform some chemicals and prevent them from reaching the fetus. Fat-soluble chemicals will diffuse more rapidly through the placenta than water-soluble chemicals. However, if the level of exposure is high enough, toxic levels of a water-soluble chemical can pass through the placenta and injure the fetus. Two examples of chemicals that do this are ethanol and cocaine. These will be further discussed in the Species Differences section below.

The mammary barrier is less distinct than the other two discussed. There is no anatomical feature that prevents chemicals from entering breast tissue or breast milk. Instead, the breast is primarily composed of fatty materials and many small blood vessels. Therefore, only fatty chemicals will tend to be distributed to the breasts. Chemicals that are very lipid-soluble and have very low water solubility, like dioxin and some polychlorinated biphenyls (PCBs; chemicals that historically were used in transformers), will be preferentially stored in fatty tissues like the breasts. If a woman is breast-feeding, the chemicals stored in the breasts could be transferred to the infant in the breast milk.

Metabolism

Relative to toxicology, the basic function of metabolism is to increase the rate of elimination of foreign chemicals (i.e., xenobiotics). The same factors that allow xenobiotics to be absorbed (e.g., fat soluble) also greatly reduce their elimination from the body. As a result, we would quickly be overwhelmed by absorbed xenobiotics without a mechanism for modifying their solubility. Metabolism involves a process of biotransformation in which the chemical structure of the xenobiotic is changed. This change increases the water solubility of the chemical, which reduces the ability of the chemical to be stored in the fat, increases its rate of filtration by the kidneys (see Excretion, below), and thereby greatly increases the rate of elimination for the chemical. This function keeps us from being overwhelmed by exposure to unwanted chemicals.

Metabolism requires the use of enzymes. These enzymes are present in the cells of many organs and tissues. The liver contains the highest concentration of these enzymes, which is consistent with its primary function as the hazardous waste site of the body. In addition, these enzymes occur at the main locations of possible chemical exposure, namely the lungs, GIT, skin, and eyes. Metabolizing enzymes are also present in the kidneys, many endocrine glands (e.g., adrenals), brain, and red blood cells. In addition, bacteria in our GIT also contain metabolizing enzymes that metabolize some chemicals before they are even absorbed. Some of these bacteria are able to metabolize things we otherwise cannot, which stresses the importance of maintaining these bacteria in our GIT.

These enzymes assist in two major types of metabolic reactions of foreign chemicals. The first type, known as Phase I biotransformation, involves only the enzyme and the foreign chemical, and slightly increases the water solubility of that chemical. The second type, known as Phase II biotransformation, requires another chemical known as a cofactor; the increase in the water solubility of the chemical caused by these reactions leads to a much higher rate of elimination.

The need for a cofactor makes a Phase II reaction more likely to be affected by saturation than Phase I reactions. In the saturation process, more chemical reaches a cell than the metabolic processes can control. In this case, the chemical builds up in excess and can cause toxic effects. Often, the amount of a cofactor present limits the rate at which these reactions can occur. One cofactor is s-adenosylmethionine (SAM), discussed in Chapter 4 in its role in detoxifying arsenic. As was mentioned then, this metabolism can be saturated if the amount of arsenic present is greater than the available SAM.

There are several dozen different types of enzymes that metabolize specific types of chemicals. For example, alcohol dehydrogenase is an enzyme that specifically metabolizes alcohol, and acetylcholinesterase is an enzyme that specifically metabolizes one of our neurotransmitters and is affected by organophosphorus insecticides (see the Selection of Toxic Endpoints section below).

One specific kind of Phase I enzyme deserves mention because of its widespread use in research, as well as the valuable information we have obtained regarding metabolism of chemicals through this research. This type of enzyme is known as cytochrome P450, or the mixed function oxidases (MFO). This type of enzyme ranks first in terms of the number and types of chemicals it can metabolize and detoxify. The highest concentration of these enzymes is found in the liver. By experimentation, we have found which chemicals stimulate and which chemicals inhibit specific enzymes within this group. This knowledge has led to increased understanding of how many toxic chemicals exert their effects, and to our growing ability to design chemicals to counteract these effects.

Usually the change to increase water solubility decreases or eliminates the toxicity of a chemical. However, occasionally a Phase I reaction will make a chemical more toxic by exposing a particular part of the chemical to a target. For example, benzo(a)pyrene is a PAH that is a probable human carcinogen. This chemical undergoes a Phase I reaction that leaves an active oxygen radical exposed to DNA. This active radical binds with DNA and causes a mutation. In the absence of this metabolism, this chemical is not toxic.

The metabolism of xenobiotics is extremely complex and occurs in diverse ways. The reader is referred to the end of this chapter for suggested further reading regarding the metabolism of foreign chemicals.

Excretion

The final component of toxicokinetics is the elimination of the chemical from the body. This is the process of excretion. The major sites of excretion in mammals are the kidneys (via urine) and the GIT (through feces). (We will look more closely at the two major sites of excretion.) Lesser methods include excretion (from the mother, not the fetus) into breast milk (already discussed), excretion through the lungs via exhalation, and excretion through sweat and saliva.

Generally, foreign chemicals are not excreted through the lungs, and only gases will be released during exhalation. This is the technical basis behind the breath analyzer test for alcohol. When we exhale, alcohol vapors from our blood will be exhaled. The breath analyzer then detects this gas. Similarly, neither excretion through sweat nor excretion through saliva is typically an important route of excretion.

Excretion Through the Kidneys

The kidneys receive about 25 percent of the blood output from the heart with each beat; 20 percent of this is filtered by the glomeruli (the filters of the kidneys). This high blood flow brings a large amount of chemicals into the kidneys. Although the kidneys function to eliminate chemicals from the body, they are also designed to reabsorb nutrients and water that are filtered through the glomeruli back into the bloodstream. There is a duel going on inside the kidneys between maximizing the recycling of useful chemicals and the elimination of harmful chemicals, including those of our own metabolism such as lactic acid. Metabolic processes increase the water solubility of the chemicals. This is intended to increase the chance that the chemical will reach the kidneys and be filtered, and to decrease the likelihood that the chemical will be reabsorbed into the blood. Damage to the kidneys will decrease their efficiency in eliminating harmful chemicals as well as decrease the reabsorption of useful chemicals. This is why dialysis is necessary if kidney function is severely impacted. Without dialysis, the amount of harmful chemicals in the blood would harm the person. Dialysis filters out the unwanted chemicals, keeping concentrations below harmful levels.

Excretion Through the GIT

Excretion through the GIT is in the form of feces. This can include undigested foodstuffs, chemicals secreted in bile into the GIT from the liver, and transfer of chemicals from the blood into the GIT (opposite of absorption). The latter of these is usually a minor pathway of excretion because it depends on diffusion and is a very slow process. One example of the first type of GIT excretion is that paraquat, discussed earlier in its use as a herbicide on marijuana and effects on the lungs, is very poorly absorbed via ingestion. Most of a dose can be found in the feces, never having been absorbed. This further indicates that paraquat is extremely toxic to humans. Very little is absorbed from the GIT, yet swallowing less than 1 teaspoon could kill you. This means that very little paraquat is actually reaching the lungs, and yet it still can have a toxic effect.

The nature of the chemical can impact the amount that is excreted unchanged through the GIT. For example, Vitamin C is often swallowed as a pill; this Vitamin C is less well absorbed than Vitamin C that is contained in fruits and vegetables. More of the chemical will be excreted unchanged when consumed in pill form because the Vitamin C is stuck to particles. Vitamin C contained in fruits and vegetables is not as bound up in a matrix that prevents absorption. Vitamin C in pill form is less likely to interact with the mucous lining the GIT due to the geometry of the particles comprising the pill. Therefore, more of the chemical will be excreted without being absorbed. This effect also holds for chemicals stuck to soil particles; absorption will be lower for chemicals stuck to soil particles than if the same amount of the chemical were dissolved in liquid and consumed. Because of the lipophilic nature of many chemicals, they are often better absorbed if dissolved in oil (e.g., corn oil) than in water.

Excretion through the bile is an important mechanism used by the liver in chemical elimination through the GIT. While much of the specific mechanistic details are still not understood, it is clear that the liver can transport chemicals from the bloodstream into the bile, and from there into the intestine. We know that there are multiple active transport systems whereby the liver can move chemicals into bile from blood. Therefore, chemicals excreted through the liver will be eliminated from the body through feces; chemicals excreted through the kidneys will be eliminated from the body through urine.

Species Differences

As we have seen from the discussion of toxicokinetics, the anatomy and physiology of an organism play key roles in defining the toxicity of a chemical to that organism. Birds have feathers rather than hair; this impacts absorption. Aquatic mammals like muskrats have an insulating layer of fur that prevents water from reaching the skin. This will also affect absorption. Cattle have four stomachs and have a large amount of various bacteria in their stomachs. This impacts both absorption across the gut and metabolism of the chemical by bacteria, even before it is absorbed. Chocolate is toxic to dogs because they lack an enzyme that metabolizes it so that it can be excreted. Differences in metabolic processes across species play a primary role in impacting the sensitivity of various species to chemicals.

Finally, physiological and anatomical differences can impact excretion. For example, the kangaroo rat lives in the desert yet requires no water to survive because its kidneys are extremely efficient at reabsorbing chemicals. However, this could also lead to increased retention of toxic chemicals, which could increase the toxicity of a chemical relative to other species. These issues are more specifically discussed below.

Genetics

Each person's genetic composition varies from all other humans (unless you are an identical twin), and also from all other organisms. It is not surprising that these genetic differences affect the toxicity of chemicals to different species. Some genetic-based differences are obvious. For example, the pesticide industry targets chemicals at specific species that are more sensitive to the toxic effects of the chemicals than other species. For example, herbicides are used to kill weeds, yet have little effect on trees or humans at the doses that are toxic to these weeds. We have developed chemicals that specifically target spiders (acaricides), fungi (fungicides), mollusks (e.g., snails; molluscicides), roundworms (i.e., nematodes; nematocides); insects (insecticides), and rodents (e.g., rats; rodenticides). Except for rodents, none of these target species are mammals like us. It is not surprising, therefore, that we are less sensitive to these other chemicals than we are to rodenticides. The relative potency of these different classes of pesticides to humans is lowest in herbicides, then progressively increases from fungicides to molluscicides, acaricides, nemato-cides, insecticides, and tops out at rodenticides. This parallels the relative genetic similarity of humans to these other target groups (genetically, we are most different from plants and least different from rodents).

Even within a single species, genetic differences can have profound effects on rela-tive chemical toxicity. For example, the drug iproniazid, which is a monoamine oxi-dase (MAO) inhibitor that used to be given as an antidepressant, is metabolized in the liver by an enzyme known as acetyltransferase. If the levels of this enzyme in the liver are lowered, the drug will be metabolized more slowly and will remain active in the body for a longer period of time. As a result, the drug will accumulate in the liver of people with slower metabolisms, even at typically prescribed doses. The result can be liver toxicity. In other words, a therapeutic dose for one person could lead to liver injury for another person. These two people could be identical in all ways except that one has a lower level of this enzyme in the liver.

The amount of this enzyme present in the liver is genetically determined. In some populations, as in Japanese and Eskimo populations, the level is almost never af-fected—the gene responsible for the slower metabolism is almost absent. However, in other populations, as in African populations, up to 80 percent (four of every five!) have lower levels of this enzyme and could be affected in situations where it is needed to destroy the chemicals in the body. (It should be noted that iproniazid, the example chemical used above, is no longer available in the United States be-cause of other types of toxic side effects it has had on people. However, acetyltransferase is an enzyme that acts on many different chemicals that are still in use in drug therapy and released into the environment.)

Kinetics/Metabolism

As we have seen in the previous example, the rate of metabolism can impact chemi-cal toxicity. Many other factors can impact the rates of absorption and excretion, or even the distribution of the chemical within the body. For example, absorption through the skin is different across species. Some species have fur or feathers, which may protect against chemicals reaching the skin to be absorbed. Others like insects have exoskeletons, which are essentially coats of armor against absorption. How-ever, insects have tiny holes in their exoskeletons called spiracles through which they "breathe." As a result, chemicals are often more quickly and extensively ab-sorbed in insects than they are in humans. This is why ant sprays are so effective when the chemicals in the spray contact the ant.

Differences in excretion rates also can impact chemical toxicity. For example, the kangaroo rat that lives in the deserts of Arizona and New Mexico is physiologically adapted to live without water. They can get enough water from their plant food to survive. This is due to a very efficient kidney that reabsorbs essentially all fluid that passed through the filter back into the bloodstream. Along with reabsorbing water, they also might reabsorb chemicals in the water that would otherwise have been excreted. This could lead to a longer duration of chemical activity than in other mammals, which could increase the effect of chemical exposure.

Sex and Hormonal Status

Some chemicals differ in their toxicity between males and females. Both the nature and location of toxic effects could be different in each sex. The critical factor that determines if a chemical might have different toxicity across sexes is a link between the metabolism of the chemical and sex hormones. Males often metabolize chemicals more quickly than females. In most cases, this leads to more resistance in the male because metabolism usually detoxifies a chemical. In those few cases where metabolism actually increases the toxicity of a chemical (e.g., cancer effects of PAHs), males may be more sensitive to the chemical than females.

Changes to hormone status other than sex hormones (e.g., hyperthyroidism) could increase sensitivity to a chemical. Too much thyroid activity can interfere with the male sex hormone actions on metabolism of chemicals. As a result, the benefit of faster metabolism is offset by the action of the thyroid.

Chemicals that affect hormone status are referred to as endocrine disruptors. This recent area of study has revealed new ways of evaluating the toxic potential of chemicals. These types of effects may be the underlying mechanism behind much of the toxicity seen for a wide range of chemicals (e.g., DDT). This concept will be further discussed in Chapter 10.

Nutrition and Dietary Factors

Deficiency of essential nutrients or vitamins can directly cause toxic effects. For example, lack of enough Vitamin A reduces the ability of the MFO enzyme system to metabolize chemicals, which has the effect of making most chemicals more toxic. Lack of available cofactors that are necessary in order for some metabolic reactions to occur could also limit the amount of metabolic detoxification. This could also result from a low protein diet. Also, poor nutrition might lead to mobilization of fat, which could release stored amounts of chemicals and lead to toxic effects.

Some essential elements, like zinc, can protect against the absorption of toxic metals like lead when they are present in the GIT. This is because the two elements compete for absorption, and zinc is favored over lead by the active processes that occur in our intestine.

Age

It is generally agreed that the very young and the elderly are more sensitive to the toxic effects of chemicals. Many specific systems are not fully mature or developed in infants. This includes the blood-brain barrier, enzymes that metabolize alcohol, and much of the P-450 system of enzymes primarily responsible for metabolizing foreign chemicals. Because the defense mechanisms against chemicals are still not complete, chemical levels that would not lead to toxicity in adults could affect small children. One example is alcohol, which the infant cannot metabolize because the enzyme that degrades it, alcohol dehydrogenase, has not yet been produced. As a result, alcohol ingested by the mother during pregnancy can accumulate in the fetus, and the baby could be born with alcohol distress syndrome (i.e., alcohol poisoning). Another similar example is cocaine, which can reach the brain in high levels because the blood-brain barrier is not yet fully developed. As a result, cocaine can be present in infants from use by the mother during pregnancy at high enough levels for the baby to be physically addicted to the drug at birth.

There is a large increase in enzyme production and activity at birth, especially in the liver. However, even after birth, small children are more sensitive than adults to many chemicals because the defense mechanisms do not mature quickly. This is why some drugs taken by adults should be given to a small child only in smaller doses (e.g., ibuprofen). Ibuprofen is stored in the body almost exclusively in plasma proteins. These proteins are not fully developed in infants. Because less ibuprofen is stored in the proteins, more is available to act in the body. This increases the effective dose a small child receives as compared with an adult. It is not until children are several years old that all of their chemical defenses are fully developed.

For the elderly, this increased sensitivity is likely due to the reduced activity of cells in general and enzymes in particular, which reduces the immunity of elderly people against both infections and chemicals. The metabolic efficiency of the liver is less in the elderly, so chemicals are detoxified more slowly. Also, the number of storage compartments where chemicals may be retained without toxicity may be lower in the elderly. For example, bone tissue becomes more brittle with age, which is in part due to loss of bone cells. Therefore, chemicals that are typically stored in bone, such as lead, might re-enter the circulation and lead to toxicity in the elderly. This potential becomes higher if they are also exposed to lead in the environment in their later years. What might be a safe dose for a healthy adult might not be safe for that same adult thirty years later.

In Chapter 7 we will discuss more fully how all of these variables are handled in developing "safe" concentrations for humans.

Physiology

Cattle and some other species (e.g., sheep) that have a high percentage of grass in their diet have multiple stomachs. These function like a series of stomachs in humans; food is broken down by hydrochloric acid and enzymes to enhance absorption in the intestine. Grass is very difficult to digest. With our single stomach, humans cannot digest much cellulose, which is a primary component of grass cells. Bacteria that live in our intestines are able to digest (or metabolize) some of the cellulose we eat and allow some of the breakdown products and nutrients to be absorbed. In cattle, however, different types of bacteria live in the multiple stomachs, as well as in the intestine. These bacteria specifically digest cellulose, allowing cows to absorb a much higher percentage of their food. Having multiple stomachs, combined with the practice of regurgitating their food (i.e., chewing their cud), aids in digesting the plant cells that comprise their diet.

Toxicologically, these differences mean that the cow will likely have more metabolism occur in its GIT than a human. Chemicals that are detoxified through metabolism, such as occurs by bacteria in the GIT, should be less toxic in cattle than in humans. However, the multiple stomachs and longer intestine means that the chemicals will be present in the GIT for a longer period of time than they would be in humans; this would tend to increase absorption of chemicals. These competing factors represent a simple example of the difficulty in extrapolating toxicity and potency across species, even based just on a couple of physiological differences. In reality, many physiological factors differ across species. In addition, the multiple factors discussed previously also have a myriad of possible ramifications to the toxicity and potency of chemicals both across species and within a given species. The combination of all these competing factors on toxicity and potency is the focus of the discipline of comparative toxicology. This branch of toxicology attempts to understand and explain the variability of toxic effects and potencies across species. One goal of this effort is to provide better and more accurate predictive power in the area of toxicology so that in the future we can reduce the need for toxicity testing and build better chemicals that are low in toxicity.

Selection of Toxic Endpoints

Given this tremendous variability across and within species, how does one select an effect that can be used to measure toxicity in humans (or any other specific species)? To do this appropriately, a good understanding of toxicology is necessary. In order to select an appropriate and relevant endpoint, the end use of the information should be considered. Is the endpoint being selected to evaluate an endangered species (e.g., survival of black-footed ferret pups) or an unborn child?

Is the goal to register a new pesticide? If the chemical is a pesticide, will it be used on food crops (e.g., methyl bromide on strawberries) or other crops (e.g., cotton)? Answers to these questions may change the endpoints that should be considered.

Some endpoints are specific toxic effects. For example, a certain class of insecticides known as organophosphates (OPs; e.g., parathion) exert their toxicity in humans by inhibiting the action of the enzyme acetylcholinesterase. This enzyme destroys and ends the activity of one of our neurotransmitters (acetylcholine) that functions to relay signals across cells. If only a small percentage of the enzyme's activity is inhibited, there is no toxicity. However, if a higher percentage (perhaps 50 percent or more) of the enzyme's activity is inhibited, then neurological toxicity can result because the acetylcholine continues to act after the signal has been relayed. This makes these nerve cells continue to relay messages even though there is no signal to send. The toxicity that results from this continuous action of specific nerve cells focuses on certain types of nerve fibers and can include increased salivation, tearing, perspiration, nausea and vomiting, urinary incontinence, wheezing, or even paralysis or coma. This demonstrates that certain effects are precursors to other, more serious effects. These types of effects are known as biomarkers.

Some effects, such as the OP insecticides inhibiting acetylcholinesterase, are relatively consistent across a large group of organisms; these can be used without much uncertainty to assess potential toxicity to a target species. Indeed, we use the results of OP studies in chickens to identify a target level of enzyme inhibition in humans that might lead to toxic effects.

Many times, rather than selecting a specific effect, maximum concentrations that cause no effects are used. In this way, it can be ensured that toxic effects will not occur as long as allowable concentrations remain below this threshold. These issues are further discussed in Chapter 7, which explains how such concepts are used to establish "safe" levels of exposure in humans.

Additional Reading

Hodgson, E. and F. E. Guthrie. 1984. *Introduction to Biochemical Toxicology*. Elsevier Science Publishing Co., Inc. New York, NY. 437 pp.

7

Extrapolation of Toxicity Values from Animals to Humans

Most of our toxicity information comes from research on animals, especially laboratory animals like guinea pigs. As we discovered in Chapter 6, there are many factors that influence the toxicity of chemicals between species, and even within a species. For example, PAHs have been shown to cause skin cancers in laboratory rodents. These tumors result from what are known as "skin painting" studies (see Chapter 5) where the skin is shaved, and the chemical of interest is directly applied to the skin. Considering the multiple genetic, anatomical, and physiological differences between rodents and humans, there is a great deal of uncertainty in applying the results of these skin painting studies to humans. Also, the type of chemical application used in the animal studies does not represent how a person could be exposed through the skin because we would not paint PAHs on our skin. How then do we establish "safe" chemical concentrations to protect humans from chemical toxicity?

To use toxicity information on animals to set guidelines and regulations, as well as to evaluate health hazards in humans, data collected on animals is often adjusted to "equivalent" levels in humans. This estimation of values in one species (humans) from data on another species is referred to as extrapolation. This chapter discusses how this process works, and what it means to an average consumer. In addition, toxicity in other animals is also expected to vary amongst species. If we want to establish a safe concentration of PAHs in soil to protect bald eagles from skin exposure, the results of skin painting studies on rodents would also need to be extrapolated because we shouldn't expect the same sensitivity for such different animals. The extrapolation process for use in wildlife species will also be discussed in this chapter.

Identification of Appropriate Animal Study for Extrapolation to Humans

The first step in adjusting the results of animal toxicology studies for humans is selecting the animal study (or studies) that will provide the raw data. We learned in Chapter 6 that there is wide variability of chemical toxicity across species, and also variability within a species. In most instances, there are numerous differences between the tested animal and the target animal (i.e., humans). Experience with countless toxicology studies in numerous animals has given us the ability to somewhat accurately predict the differences across species.

Differences in physiology, especially in enzyme activity, account for a significant amount of this variability. For example, the chemical 6-propylthiopurine, a drug used to treat hyperthyroidism, is metabolized through a specific pathway in the mouse to a potent carcinogen. However, humans metabolize the chemical differently, and the metabolic products (known as metabolites) do not cause cancer. In this case, using the mouse data to set "safe" levels in humans would greatly overestimate the toxicity of the chemical to humans.

For other chemicals, the opposite can be true. For example, the chemical 2-acetylaminofluorene (2-AAF), which is used in liver toxicology research, is activated by the P-450 enzyme system in the liver to a carcinogen in humans but not in guinea pigs because we metabolize the chemical differently than do guinea pigs.

Similarly for wildlife species, differences in physiology have dramatic effects in toxic sensitivity. For example, although peregrine falcons are very sensitive to the eggshell thinning effects of DDT (Chapter 2), other species of birds are not impacted. For example, this effect has not been observed in game birds (e.g., chickens, turkeys, quail, pheasants). In many instances, these differences are based on differences in the type, speed, and end products of metabolism across species.

In addition to these differences in metabolism, the impact of age, nutritional status, sex, and other factors also needs to be considered. The examples presented above illustrate the tremendous amount of uncertainty that lies in using results of animal toxicology studies for humans or other species. For this reason, the U.S. EPA, FDA, and international organizations like the World Health Organization (WHO) and IARC, have established standardized animal testing protocols that must be completed before introducing new chemicals into the marketplace or the environment. These protocols, as discussed in Chapter 5, include testing on multiple species and for multiple endpoints (e.g., reproduction, cancer). This increases the amount of comparative data available for a chemical, which allows us to better understand how toxicity occurs in different species. This knowledge decreases the amount of uncertainty involved in extrapolating data across species, because we better understand how the chemical works in the body and the factors that affect its toxicity.

The goal of selecting an appropriate animal species should be to minimize the uncertainty of both the nature and severity of toxic effects from chemical exposure across species. To do this, scientists "match" the animal study with the expected human exposure. This includes the concentration and duration of exposure and the type of toxic effect, as well as similarity in physiology and the other factors discussed above and in Chapter 6. If the expected human exposure is through the skin, an animal that has skin physiology similar to humans should be used (i.e., the pig). However, if the expected human exposure is through the GIT, then a monkey would provide more reliable data because its intestinal physiology is more similar to a human's than to laboratory rodents. Similarly, if humans are likely to be exposed to a chemical over several years, a long-term (or chronic) animal study would better represent expected conditions than a one-dose (e.g., acute) study.

When selecting an appropriate animal study for extrapolation to humans, the type of toxic effect measured is an important consideration. There are generally three types of toxic endpoints evaluated in laboratory animals:

- The highest dose that causes no effect (a no-observed adverse effect level; NOAEL)
- The lowest dose that causes a toxic effect (a lowest-observed adverse effect level; LOAEL)
- The dose or concentration that kills 50 percent of test animals (LD_{50} or LC_{50}; Chapter 5)

The LD_{50} and LC_{50} are least relevant for humans, because we do not want to have lethal amounts of chemicals in the environment. Such studies are typically used to establish a ceiling for doses in further experiments, which are necessary to establish the critical dose-response relationship for each chemical. If possible, we want to keep the chemical at concentrations below threshold levels. Therefore, the NOAEL is usually the most relevant endpoint for extrapolation of animal data to humans.

Most often, the appropriate animal species is based on required protocols, and typically includes a rodent (e.g., mouse) and one other species in a different phylogenetic group (e.g., a rabbit, which is not a rodent).

Conversion of Dose

After the most appropriate animal study is selected, the amount to which the test animal was exposed needs to be converted to an equivalent amount in humans (or another animal species). An amount equal to a grain of sand might be enough to affect a mouse, which weighs only about 25 grams (3/4 ounce). This same amount would be unlikely to affect an adult human, who might weight 70,000 grams (155 pounds). Therefore, more of the chemical would need to be given to an adult human to be an equivalent amount per unit weight.

Also, the amount given to a mouse might be over the course of 3 months, but the human might only get the same amount after 10 years of exposure. So, in addition to normalizing a dose by body weight, it must also be normalized for different lengths of exposure. Typically, a period of one day is used for this normalization. This amount is referred to as a dose. A dose is defined as the concentration of a chemical per unit weight of the organism, and is often expressed in units of milligrams of chemical per kilogram of body weight per day (mg/kg/day).

Dose conversion across species to a human is a multi-step process. Each factor is separately evaluated, and a value or equation is used to account for the possible differences. The following factors are separately evaluated:

- Differences based on species (e.g., genetic, physiologic)
- Variability within a species
- Differences in study duration between laboratory animal and target duration in humans
- Differences in exposure route between laboratory animal (e.g., intravenous) and humans (e.g., food)
- Type of endpoint in laboratory animal (e.g., LOAEL) and target effect in humans (e.g., NOAEL)

Each of these factors is discussed below.

Extrapolation across Species

As we have seen, there are many differences among species that will affect the toxicity of a chemical. Although we understand a few of these differences well enough to predict their effect on other species for some chemicals, we have no information for the vast majority of chemicals. There are approximately 90 million chemicals known; about 7,000 of these are considered important based on their usage. We have compiled reliable toxicology information and identified mechanisms of action for less than 1,000 chemicals. Usually, these mechanisms of action have been determined for animals rather than humans. Therefore, we often cannot predict with any confidence whether humans will be more or less sensitive to a chemical than the animal species on which information is available. As a result, we use an "uncertainty factor" to account for this lack of knowledge. The uncertainty factor is used to adjust the experimental dose to a more restrictive level in the target species (e.g., humans). In essence, the uncertainty factor adjusts the dose so that the chemical is assumed to be more toxic in the target species than the test species. In this way, we make it less likely that we will underestimate the toxicity of chemicals to the target species.

For extrapolation to humans, the U.S. EPA has adopted a convention of applying an uncertainty factor of 10 to the dose tested in animals. This is equivalent to assuming that the chemical is 10-times more toxic in humans than in the test animal. Although this factor was not based on science when it was developed in the 1950s, further studies have demonstrated that this factor is protective in 95 percent or more of cases.

A similar process is used for extrapolation from one animal species to another species in the environment. However, the uncertainty factor used increases as the taxonomic difference between the test and target species increases. For example, extrapolating results of a laboratory mouse study to evaluate toxicity in a ground squirrel from exposure to lead released from a smelter might only require an uncertainty factor of 2 (ground squirrels and mice are in the same taxonomic family). Extrapolating the results of this same study to a fox might require a factor of 10 because these two species are more distantly related.

In most cases, use of an uncertainty factor leads to overestimation of chemical toxicity in the target species (e.g., humans). However, there are rare cases where humans might be 10 or more times more sensitive to a chemical than the target species. In such cases, the uncertainty factor might not adequately account for the different sensitivities. However, laboratory species are typically bred to be sensitive to the toxic effects of chemicals. In addition, multiple species are tested and typically the results of the most sensitive animal tested are used as the basis for the extrapolation.

Intraspecies Variability

In addition to differences among species, there are also differences in toxicity between individuals of the same species. As previously discussed, metabolic differences between individuals makes a subgroup of the population more sensitive to the toxic effects of some chemicals than others. In addition, young children and the elderly are typically more sensitive than others. Because animal tests are typically conducted on individuals with a very narrow range of genetic variability, the results of the tests do not always indicate the range of sensitivity to toxic effects that might occur in a diverse group. As a result, an uncertainty factor is typically used to account for sensitive members of a population. Again, the goal of using an uncertainty factor is to develop a dose that will be below any threshold that could cause toxic effects in any members of the target species. The U.S. EPA also uses a factor of 10 to account for this variability. This assumes that sensitive individuals could be affected by doses 10 times less than those given to others.

Extrapolation Across Exposure Duration

In most cases, we are concerned with long-term chemical exposure. The exposure might occur in the workplace (e.g., steel worker, dry cleaner), at home or school (e.g., air emissions from a smelter or refinery), or elsewhere (e.g., diesel exhaust from trucks and buses on the freeway). In all of these situations, we might not be exposed to very high chemical levels, but we will likely be exposed intermittently for many years. This is considered chronic exposure, which is assumed to be more than 10 percent of a lifetime. For humans, this includes exposure durations of seven years or longer. Most laboratory animal studies are either acute (i.e., short-term exposure to high concentrations), or subchronic (moderate-length exposure). Typically, a subchronic study duration is 90 days. This period of time represents about 10 percent of the lifetime of a rat, but less than 10 percent in species such as rabbits, dogs, or monkeys. Long-term laboratory animal studies typically range from six months to two years in duration. These are often cancer studies, which require a long latency period for the cancer effects to develop. Therefore, the studies need to be longer. Effects other than cancer are identified in these studies, which are considered chronic studies. An uncertainty factor is not needed if a chronic study is used. However, in many cases only a subchronic study is available. In such situations, an uncertainty factor is applied to account for the potential for a lower dose to cause similar effects at a higher dose if exposure occurs over a longer period of time.

For extrapolation to humans, the U.S. EPA has adopted a convention of applying an uncertainty factor of 10 to the dose tested in animals. This is equivalent to assuming that the chemical might be 10 times more toxic if exposure were chronic rather than of shorter duration. The same process and factor is typically used for extrapolation to wildlife species.

Extrapolation across Exposure Route

Sometimes the route of exposure will differ between the tested animals and the target species. For example, most studies on chloroform were via the inhalation route of exposure because of chloroform's historical use as an anesthetic. This chemical is quite volatile, and vapor inhalation was likely to represent the major route of exposure in most situations. However, chloroform released into the environment has often impacted drinking water aquifers. In this situation, exposure would be through drinking water, not inhalation. We discussed previously that the rate of absorption, nature of toxic effects, and sensitivity to effects often differs by exposure route. Therefore, extrapolation across exposure routes also requires use of an uncertainty factor. Using the chloroform example, we know that the toxicity of chloroform is due to metabolic action that occurs primarily in the liver. In inhalation studies, chloroform is first absorbed into the lung. When ingested, it first reaches the liver before entering the general circulatory system. This means that, assuming the percent of absorption is similar across both routes, more chloroform will reach the target organ from ingestion than from inhalation. Using this argument, chloroform could cause toxic effects at lower doses from ingestion than from inhalation.

Rather than using a fixed factor of 10 to account for this uncertainty, U.S. EPA typically assumes that toxicity across different routes of exposure is the same. Therefore, a factor of 1 is usually used.

Additionally, in some situations extrapolation across exposure routes is not recommended at all. For example, we know that metals often have different target organs depending on how exposure occurred (e.g., breathing, eating, or drinking). For metals where we do not have toxicity data for both exposure routes, there is no way to accurately predict if toxicity will be the same or different. Therefore, extrapolation across exposure routes is not typically performed for metals. For most environmentally important metals, toxicity information has been compiled for multiple routes of exposure, which obviates the need for such extrapolation. The same process is typically used for extrapolation to wildlife species.

Extrapolation across Effects

For extrapolation to humans, the U.S. EPA has adopted a convention of applying an uncertainty factor of 10 to the dose tested in animals. This is equivalent to assuming that the chemical might have an effect at a lower dose than that used in the study. The difference in dose between having a minimal effect and no effect is assumed to be a factor of 10. This is illustrated in Figure 7.1. Assuming the dose-response relationship for a chemical is linear, the percent of individuals reporting toxic effects declines as the dose is lowered. The factor of 10 assumes that the slope of the line is such that there is a 10-fold difference between a LOAEL and NOAEL (Chemical A). However, if the slope of the line is different than this, the difference between a LOAEL and NOAEL might be greater than 10 (Chemical B) or less than 10 (Chemical C). The relationship between the LOAEL and NOAEL for a given chemical is based on how the chemical exerts its toxicity in the body (i.e., its mechanism of action). For most chemicals, because we do not understand the mechanism of toxicity, especially at doses lower than those typically tested in the laboratory, this represents a large level of uncertainty in dose extrapolation.

Many studies report both a NOAEL and a LOAEL in a given laboratory species. The results of these studies can be used to calculate an actual factor that relates a LOAEL to a NOAEL within a species. For most chemicals, the ratio between a LOAEL and NOAEL is a factor of five or less. Therefore, the factor of 10 used is adequately protective for the vast majority of chemicals. It is only chemicals with very flat dose-response curves where the difference between a NOAEL and LOAEL could be more than a factor of 10. A very flat dose-response curve means that a large increase in the dose results in only a very small increase in the frequency of toxicity. These chemicals (e.g., toluene) are typically less dangerous than those with steep dose-response curves because a small difference in the dose does not have dramatic consequences in toxicity. For example, cyanide is a dangerous chemical because it has a very steep dose-response curve; once toxic levels are reached, death can result from only a small increase in the amount.

The same process and factor is typically used for extrapolation to wildlife species.

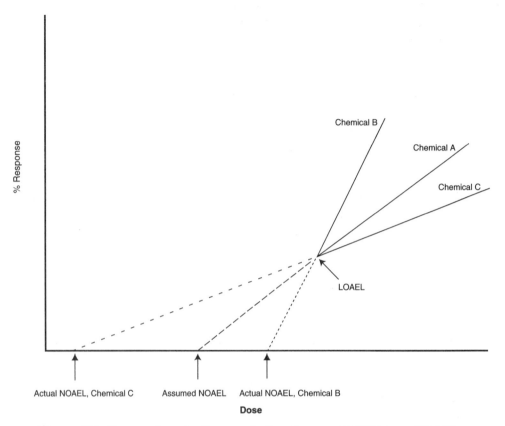

Figure 7.1. Uncertainty in Extrapolating from a LAOEL to a NOAEL

Examples of the Dose Extrapolation Process

Due in large part to all of the uncertainties associated with this process, the U.S. EPA, WHO, and other regulatory bodies have developed a standardized set of uncertainty factors that are used to conduct extrapolation of toxicity values across species:

- Factor of 10 to extrapolate from another species to humans
- Factor of 10 to extrapolate from an average human to a sensitive human
- Factor of 10 to extrapolate from anything less than a long-term study to a chronic basis
- Factor of 10 to extrapolate from a level causing an effect to one where no effect would occur

In addition, a modifying factor can also be used to further influence a value. A modifying factor might be used if there was only minimal data available for extrapolation. This could lead to greater uncertainty because, for example, multiple species were not tested. Because uncertainty might be greater, the values of 10 used for each factor might not sufficiently account for variability. Use of modifying factors is rare; typically values of 2 or 3 are used when appropriate. In all cases, these factors are used to lower the equivalent human dose, which has the impact of increasing the assumed toxicity of the chemical to humans relative to other species.

Ideally, extrapolation of toxicity data can be minimized by using data available for humans. However, as previously discussed, sufficient information is not often available for humans.

To illustrate the processes used for chemicals where (1) extrapolation from other species is necessary, and (2) data from humans are used, examples from U.S. EPA publications are presented below.

Example 1: Toluene

Toluene is a common solvent, and is used in model glues. Most toxicity in humans from exposure to toluene is from glue-sniffing or accidental exposure in the workplace. These incidents rarely provide information as to the amount to which someone was exposed. In this example, we are interested in ingestion of toluene. For example, take a case where toluene-contaminated groundwater was used as a drinking water supply and we want to be sure that the toluene levels are below those that could cause toxic effects in humans. There is not sufficient data for humans to establish a safe concentration or dose for this type of exposure.

Instead, a study conducted on rats was used and extrapolated to humans. The rat study was a subchronic study conducted by the National Toxicology Program and lasted 13 weeks (91 days). Toluene was administered to the rats via gavage. Both a NOAEL and LOAEL were determined in the study. The NOAEL was 223 mg/kg/day. The LOAEL was 446 mg/kg/day. The reported toxic effects at this dose were changes in the weight of both the liver and kidney in exposed animals. At higher doses, toxic effects in the cells of the liver and kidneys were observed. Therefore, the LOAEL endpoint of different organ weights provides an early warning sign of toxicity. These warning signs of toxicity are called biomarkers.

Three of the four factors listed above are relevant to this study. Extrapolation across species, variability within a species, and from a shorter duration to a chronic duration need to be conducted. For this study, no extrapolation from a LOAEL to a NOAEL is necessary because a NOAEL is provided. Note however that the difference between the NOAEL and LOAEL in this study was only a factor of two. This is five-fold less than the factor of 10 assumed by U.S. EPA for chemicals in the absence of a NOAEL.

Each of these three factors of 10 is used to adjust the NOAEL dose reported in the mice (see Table 7.1). All of the factors are multiplied together to result in an overall uncertainty factor of 1,000 (10 x 10 x 10). The NOAEL dose of 223 mg/kg/day is then divided by this total uncertainty factor of 1,000, resulting in a dose of 0.223 mg/kg/day. This is rounded to 0.2 mg/kg/day by U.S. EPA.

To understand what this means, realize that the original measured dose of 223 mg/kg/day daily for 91 days caused no effect in rats, even when the chemical was directly placed in the stomach. Yet the U.S. EPA regulates the chemical in the environment so that dose levels in humans should not exceed 0.2 mg/kg/day, a value 1,000 times lower than that reported in rats. If humans were exposed to 1 mg/kg/day, would toxicity result? It is unlikely because this dose level is still over 200 times lower than the no-effect dose seen in rats. However, the chemical would be considered to present a possible human health threat because the dose exceeded the conservative 0.2 mg/kg/day dose level.

Table 7.1. Animal to Human Noncancer Dose Extrapolation for Toluene

Parameter	Symbol	Value	Units
LOAEL in rats	LOAELr	446	mg/kg/day
NOAEL in rats	NOAELr	223	mg/kg/day
Interspecies UF	UF1	10	None
Intraspecies UF	UF2	10	None
Exposure Duration UF	UF3	10	None
Toxic Endpoint UF	UF4	1	None
Total UF[a]	UFt	1000	None
NOAEL in humans[b]	NOAELh	0.223	mg/kg/day

mg/kg/day = milligrams of chemical per kilogram of body weight per day.
LOAEL = lowest observed adverse effect level.
NOAEL = no-observed adverse effect level.
UF = Uncertainty factor.

[a] Total UF = UF1 x UF2 x UF3 x UF4 (10 x 10 x 10 x 1).
[b] NOAEL in humans = NOAELr/UFt (223/1000).

Example 2: Arsenic

Compare and contrast the methods and values used for toluene with those used for arsenic. We have sufficient information from arsenic contamination of drinking water in Taiwan and Chile to develop a protective dose for humans ingesting arsenic in drinking water; therefore we do not need to extrapolate results from another animal species. Both a LOAEL and NOAEL are available from the drinking water studies from Taiwan. A NOAEL of 0.0008 mg/kg/day and a LOAEL of 0.014 mg/kg/day were reported. Based on these doses, it is clear that arsenic is more toxic to humans than toluene because the arsenic doses needed to cause toxic effects are much lower than those causing toxic effects by toluene.

The only potentially relevant uncertainty factor for this study is for extrapolation to sensitive members of a population. Although more than 28,000 people were exposed to arsenic in the Taiwanese studies, these were from a specific area and did not necessarily represent the range of sensitivities to other ethnic groups (e.g., Caucasians). As a result, the U.S. EPA used a factor of 3 for this variable. This lower factor was used instead of the typical factor of 10 because data were available for humans, and some sensitive individuals (e.g., children) were included in the studies. The NOAEL of 0.0008 mg/kg/day was divided by this factor of 3 to result in a dose of 0.0003 mg/kg/day.

Because there is a much lower uncertainty factor used for arsenic than for toluene, does this mean that we are more certain about the toxicity of arsenic in humans than the toxicity of toluene in humans? Not necessarily. The human information used in the arsenic study often did not include concentrations of arsenic in the drinking water. Therefore, extrapolations were made *within the study* to estimate a concentration and daily dose associated with effects. For example, the arsenic concentration in drinking water ranged from 0.001 mg/L (parts per million) to 0.017 mg/L, a factor of 17. The arithmetic average of these concentrations was used to calculate the NOAEL. Also, the amount of arsenic exposure from food irrigated with this water needed to be estimated. This was complicated by the absence of arsenic concentrations measured in prime dietary components grown locally, including rice and sweet potatoes. Also, the amount of water consumed daily needed to be estimated. Therefore, although only a factor of 3 was applied to the dose, many other uncertainties were addressed in calculating the NOAEL. Indeed, U.S. EPA has equal confidence in the dose estimates for both toluene and arsenic, and even feels more confident with the toluene animal study than the human study with arsenic.

Overall, this should indicate that there is much uncertainty when toxicity data are used to establish "safe" levels for human exposure. The goal is to ensure that the toxicity of a chemical is not underestimated in humans so that we will have a margin of safety in setting regulatory limits for exposure. As a result, in most cases the established doses are more restrictive than necessary to protect human health.

High-to-Low Dose Extrapolation

One main concern in extrapolating data involves estimating effects at low levels from the very high levels tested. At sufficiently low levels, the effects are so minor or rare that they likely can't be measured. At the other end is death, which is easily measured. In most situations, humans will be exposed to chemical concentrations closer to the low levels, where effects are often minor, than to those causing death. However, most laboratory studies use levels just below those at which death occurs. Therefore, the degree of toxic effect at low levels needs to be estimated from effects at higher levels.

The problems inherent in such extrapolations are illustrated in Figure 7.2. We typically collect data at a few dose levels, represented by the dots. Using mathematical equations, we then draw a straight line that bisects these data points. Once we get to the lowest dose tested ("a" in Figure 7.2) the line ends because we don't know if the relationship between dose and response seen at high concentrations is the same at low concentrations.

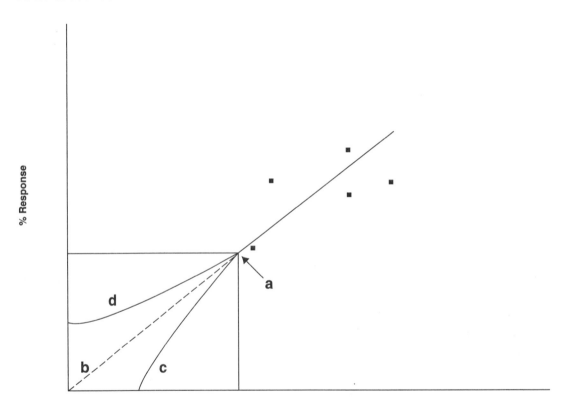

Figure 7.2. Uncertainty in High-to-Low Dose Extrapolation

If the dose-response relationship for a chemical is linear across all doses ("b" in Figure 7.2), the uncertainty associated with high-to-low dose extrapolation would be low. This relationship is assumed for all chemicals, unless information exists that indicates otherwise. For some chemicals, the effects will be lower than expected at low doses. This is referred to as a sublinear relationship ("c" in Figure

7.2). Such situations often occur when the toxicity of a chemical is based on saturating a metabolic pathway (see Chapter 5). At levels below the metabolic saturation limit, the degree of toxic response is much lower than expected based on the relationship established at high doses. In these situations, the straight-line extrapolation would lead to over-estimating toxicity of the chemical at low doses. Alternatively but less common, a chemical might have a supralinear relationship ("d" in Figure 7.2) where the effects at low doses are greater than predicted based on the relationship established at high dose levels. In these situations, the straight-line extrapolation would lead to underestimating toxicity of a chemical at low doses.

One complicating factor in this extrapolation process is that the target of interest is different for those chemicals that may cause cancer and those that don't. For non-cancer chemicals, we are interested in a threshold level, below which no toxicity occurs. For those that may cause cancer, we are interested in levels as low as possible. Therefore, extrapolation from high to low dose is more uncertain for carcinogenic chemicals, because typically the distance between the studied dose and targeted dose is much larger than it is for a non-cancer chemical. Just a little difference in the angle of the line can lead to a big difference in a regulatory value.

So far, we have only discussed dose extrapolation for non-cancer chemicals. The same uncertainties discussed for dose extrapolation of non-cancer effects are also relevant for cancer effects. Mathematical models are typically used to extrapolate relationships from dose and cancer incidence established using high laboratory doses to those at low doses. These models can range from relatively simple linear equations (e.g., a straight line is assumed to represent the relationship) to complex mathematical solutions that involve exponential terms and knowledge of calculus. These models differ in their complexity based on the amount of information known about how a chemical causes cancer. Since the process of cancer development is complex, it is not surprising that some of the models used to describe the relationships between dose and response are also complex.

Probably the most common of these is the Linearized Multistage (LMS) model, which is one of the simplest models developed for dose extrapolation of cancer effects. This model assumes the cancer potency of a chemical is linear at low doses, as shown in Figure 7.3. A best-fit linear curve is defined for data from high dose levels (the solid line in Figure 7.3), and this curve is then fitted through the origin because, in this model, no threshold is assumed for cancer effects. The shape of the curve will be uncertain due to the variability in laboratory data. For example, three sets of animals are typically used for each dose level during toxicity testing (Chapter 5). Different numbers of animals are likely to develop tumors in each group, even though the dose is the same. This is expected due to differences among individuals, even those genetically bred. This uncertainty is captured in the model by identifying confidence limits that put statistically based boundaries on the actual shape. Typically, a 95 percent upper confidence limit (UCL) is used.

Use of the 95 percent UCL value means that the actual potency of a chemical is likely to be lower than or equal to the value 95 percent of the time. In only 5 of 100 cases would the potency of the chemical be underestimated. In this way, use of the slope factor will usually lead to an overestimation of the true cancer potency of a chemical, assuming the LMS model accurately captures the shape of the dose-response curve at low doses. The 95UCL in this example is shown as the dotted line in Figure 7.3.

The slope of the 95UCL curve is then used by U.S. EPA as the slope factor to set "safe" concentrations for human exposure.

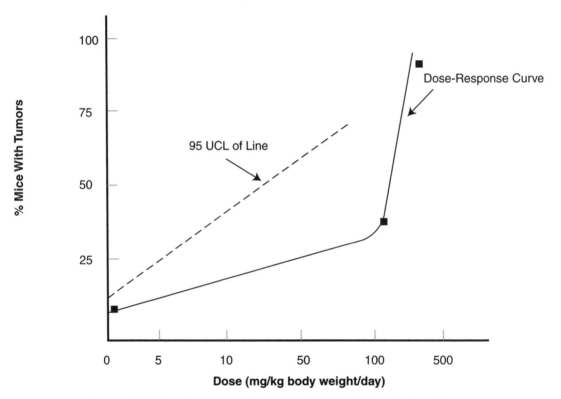

Figure 7.3. Development of a Slope Factor Using Animal Data

8

Risk Assessment

So far, we have discussed various aspects of toxicology. This discipline identifies the manner in which chemicals exert toxicity, and the potency of chemicals on various species. The majority of toxicology studies are conducted under controlled conditions in the laboratory. This is necessary to establish cause and effect relationships and to develop dose-response information on specific chemicals. However, as discussed in Chapter 7, humans are not typically exposed to concentrations tested in these laboratory studies. We learned about the uncertainty in trying to extrapolate toxicity information to humans or other species. In spite of this uncertainty, we are ultimately concerned with the potential impact of chemicals released into the environment. This issue concerns all of us because of the myriad ways we might interact with these chemicals. They can be present in our water, air, soil, or food. Estimating the likelihood of toxicity from exposure to chemicals in the environment is the focus of the discipline of risk assessment.

There is risk associated with every aspect of our lives. As long as we live, there is a finite risk that harm could come from everyday activities. For example, we could be involved in a car accident, or get struck by lightning, or get skin cancer from ultraviolet light exposure. These risks might differ from one another, but each of them could happen. However, most of us don't avoid the sun because of the possibility of getting skin cancer. Similarly, we don't avoid driving just because we could get in an accident. Even if we decided to walk everywhere, there is a risk of tripping and injuring ourselves, or getting hit by a car while walking.

The same is true for chemicals. There is a risk of toxicity from exposure to any amount of a chemical. This risk might be so large that we can be fairly certain toxicity would result (e.g., carbon monoxide poisoning from running your car in an enclosed space for a long period of time), or so small that it is essentially unmeasurable (e.g., one molecule of toluene in a reservoir). Part of the risk assessment discipline is to identify what risks are so small that we can ignore them. We will discuss this in more detail in Chapter 9.

Risk assessment impacts many different areas and is the most relevant area of toxicology for the average person to understand. For example, in certain parts of the country, regulatory agencies announce "spare the air" days because the degree of air pollution is expected to be higher than normal, most often due to weather patterns. How does an agency know when to announce a spare the air day, and what does it mean to those breathing the air?

There are occasional accidental releases of chemicals into air from industrial plants (e.g., oil refineries). Under what circumstances should we be concerned for our health when releases occur? What about industrial accidents (e.g., oil spills, train accidents and spillage)? How much chemical needs to be released in such an accident to impact health?

Additives are often used in foods, and pesticides are used to help control pests on crops. How do we know how much of these substances can be in food before we should be concerned? Should we only eat food organically grown without pesticides? Chemicals can also be in our drinking water, such as MTBE (Chapter 2). Should we be concerned about this? Will it affect our health? Should we filter our water, or drink only bottled water? How safe is bottled water, which is usually processed?

The U.S. EPA has identified some historically contaminated areas as Superfund sites. What does this mean for someone living near a Superfund site? Recently, redevelopment agencies have begun to redevelop properties that have long been unused because of chemical contamination. These are known as Brownsfield sites. How much cleanup is needed at these sites before it is safe to use the property? It is safe to live there, or to work there?

In many cases, an environmental assessment will be conducted on a property by a prospective buyer to see if the site has chemicals from previous activities at the site. How does such a prospective buyer know if the chemicals present at the site can harm people or other species? Does it make a difference if the person wants to build single-family houses at the property, or a strip mall, or an office building? What if the land is to be used for a golf course?

Risk assessment is the mechanism by which these questions are addressed, although it relies on many other disciplines to answer them. For example, environmental chemists are needed to consider how chemicals will move once they enter the environment, and how long they will last in the environment before they degrade. Geologists and hydrogeologists are needed to compile information that the environmental chemist can use to estimate the rate and degree of chemical movement in the environment.

For example, let's assume that 100 gallons of a chemical are spilled into a river. The chemical will mix with river water and move downstream. Some of the chemical will stick to the sediments in the river bottom. Does this present a risk? If so, to whom and to what extent?

Now if we assume the same spill occurs on a highway, are the risks the same? Probably not, because it is very unlikely that the chemical will move as far without flowing water providing a mechanism for moving it. However, mixing the chemical with water in the river has the affect of diluting the concentration of the chemical in the river. The same dilution will not occur in the highway spill. If this same amount of chemical were spilled onto bare soil, some could migrate through the soil and enter groundwater, which may be used as a drinking water supply for someone a mile away. How do you compare the risks in these situations? Which is worse? How much chemical do we need to clean up in each situation? This introduces an economic component into risk assessment.

Ideally, we would like to clean up every molecule that enters our environment. Due to limits in technology this is usually not possible. Let's assume that it costs $1 million to clean up all 100 gallons of the chemical spilled in the above examples. In a typical situation, it might cost $100,000 to recover 95 gallons of this total. It might cost nine times that much to recover the last 5 gallons. Using simple arithmetic, this means that 95 percent of the spill can be cleaned up for about $1,000 per gallon. The cost associated with cleaning up the last five gallons goes up to $180,000 per gallon. Using this example, it is 180 times more expensive to completely remove the chemical than it is to clean up most of it. Risk assessment is often used in such situations to identify how much chemical needs to be cleaned up to protect human and ecological health. As previously discussed, most chemicals do not cause toxicity below a certain level. Also, exposure to the remaining chemicals can be controlled. Therefore, there is no need to completely remove a chemical released into the environment because the very low levels left in the environment might not present a risk to our health. Risk assessment defines these "safe" levels.

A complicating factor is that we are typically exposed to many chemicals at the same time. If only a little bit of one chemical is left, does that present a risk when combined with the little bits of hundreds of other chemicals left in the environment? To address this question, the issue of multiple chemical exposure needs to be considered.

The overall goal of risk assessment is to establish chemical concentrations that can be present in the environment without putting humans or ecological species at risk. These concentrations will be different depending on the situation. There are two branches of risk assessment related to chemical exposure: human health and ecological. Although the overall goals are the same, there are differences in how the process is conducted. We will discuss human health risk assessment first, then ecological risk assessment.

Human Health Risk Assessment

As we have seen, there are a great variety of situations for which risk assessment can be applied. Regardless of the situation, however, the goals and objectives of a risk assessment are generally the same: to identify if either (1) an area containing chemicals may be toxic to humans, or (2) a specific chemical is toxic, and if so at what concentration? If a chemical might be toxic at the levels present at a site, then a third goal would be to identify how much cleanup is necessary to protect human health.

Because the goals and objectives are similar across all risk assessments, the general methodology has been standardized in the United States by the U.S. EPA. The methods were originally developed by the National Academy of Science in the 1970s and refined for use in the Superfund program in the 1980s. Risk assessment has been similarly standardized in other regions, including Europe. Standardization of methods ensures that risk assessments will be generally consistent regardless of the situation. This ensures that the important factors are adequately considered when assessing the potential for chemical toxicity to result from exposure.

The standardized human health risk assessment methodology generally includes the following four components:

- Data collection and evaluation
- Exposure assessment
- Toxicity assessment
- Risk characterization

An overview of the methodology and the interconnections among the components is shown in Figure 8.1. In data collection and evaluation, the conditions at the site are identified. In this step, we use information about the chemicals present at the site to identify concentrations to which we could be exposed and the spatial extent of contamination. In exposure assessment, we identify how humans could contact chemicals, and estimate possible doses resulting from assumed levels of exposure. This includes identifying the types of people that could be exposed, and by what routes they could be exposed (e.g., ingestion, inhalation). For example, would someone be exposed to lead if it was buried in soil under concrete? If the impacted area were an active industrial facility, would you assume children could be exposed? In toxicity assessment, we use the methods outlined in Chapter 7 to develop toxicity values to define the potency of a chemical to humans. In risk characterization, we combine the dose estimates and toxicity values to generate a risk value for each chemical. These risk values are compared with target risk values that are typically identified by regulatory agencies to define "acceptable" risk. Although the process is formalized, the specifics will change depending on the conditions that are relevant for a given site and situation. These four components are separately discussed below.

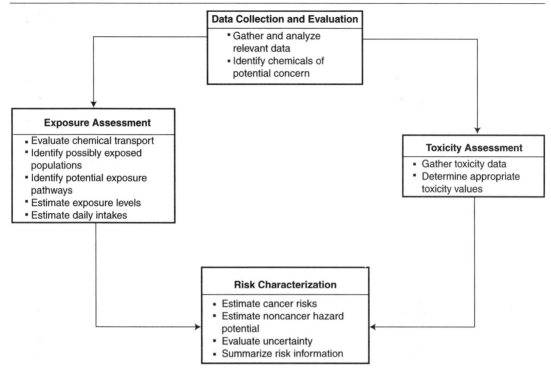

Figure 8.1. Overview of Human Health Risk Assessment Components and Process
[Adapted from U.S. EPA, 1989]

Data Collection and Evaluation

The objective of the data collection and evaluation component of a human health risk assessment is to define the nature and extent of chemical contamination at a site. This could include taking soil and/or groundwater samples to identify what chemicals are present, where they are present, and at what concentrations. This information provides the foundation for the risk assessment.

As a typical example, assume a gasoline underground storage tank at a gas station has leaked into the surrounding soil. This soil is only one foot away from a source of groundwater. To adequately address possible risks at the site, we should take samples both in soil and groundwater and analyze them for chemicals related to gasoline. Typically, this would include benzene, toluene, ethylbenzene, and xylenes. These chemicals represent the majority of the toxic ingredients in gasoline. Additives such as MTBE might also need to be looked for if they are used in the gasoline. If we looked only at the soil, we might underestimate exposure because the chemicals might have moved into groundwater that someone could drink if they had a nearby well. Drinking groundwater would expose the individual to more of the chemical, which would lead to higher exposures than people exposed only to soil. Taking groundwater samples will provide data that can be used to account for this possibility.

We could also collect air samples to see if any of the chemicals have migrated up through the soil into air through a process called volatilization. However, using the above example we would also likely need to collect background air samples to determine if the measured levels are due to our leak, the exhaust of the vehicles at the gas station, or the same chemicals volatilizing from the pumps during use.

Although it is true that someone at the gas station would be exposed to both background levels of chemicals in addition to chemical levels coming from the leak, one of the goals of a risk assessment is to identify to what level chemicals at a site need to be cleaned up to protect human health. We typically do not clean up background amounts of chemicals. Therefore, although a risk assessment may consider exposure to both background and site-specific chemical concentrations, cleanup decisions are most appropriately based on the concentrations above background.

The data collected are then usually combined statistically to generate a concentration that represents the chemical at the site. This often is the maximum concentration found at the site. Although this overestimates the level of the chemical present, it ensures that we will not underestimate possible exposures. Therefore, we are unlikely to conclude that levels are safe if they actually are not.

At the end of this component, we will have a concentration of each chemical detected in each medium (e.g., soil and groundwater). These concentrations are used in the next step to estimate doses.

Exposure Assessment

As discussed in Chapter 3, there are no toxic effects without exposure. This portion of a risk assessment is critical because it defines the conditions under which exposure is assumed to occur. Using these conditions, exposure pathways are identified and doses are estimated for each pathway. The combination of conditions and exposure pathways identified for a site is often referred to as an exposure scenario. The conditions should match those relevant at a given site.

In this portion of a risk assessment, all the ways that chemicals could move in the environment, as well as all types of people that might be located in areas where the chemicals could be present, need to be considered. Even though the leak for the gas station example introduced in the Data Evaluation section has occurred in soil below the ground, exposure could occur off-site because chemicals could be in groundwater, which moves over time. This and other transport mechanisms are illustrated in Figure 8.2. It is beyond the scope of this book to discuss the mechanisms by which chemicals can move in the environment. However, it is important to know that there are predictable factors that can be used to identify the likely ways by which people could come into contact with chemicals from a site.

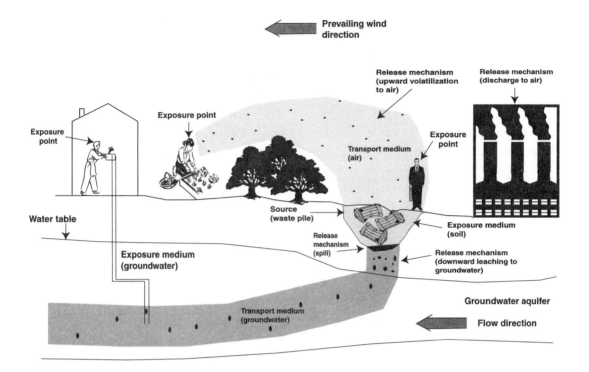

Figure 8.2. Chemical Transport Mechanisms and Exposure Pathways
[Adapted from U.S. EPA, 1989]

There are three basic steps to an exposure assessment:

- Identify human populations that might be exposed (i.e., receptors)
- Identify complete exposure pathways
- Estimate doses

Identify Human Receptors

Not everyone will be exposed to chemicals associated with a site. In this step, the types of people that could be exposed are identified. Using the gas station example, the gas station attendants will represent the people most likely to be exposed to chemicals at the site. People refueling their cars will also be exposed. Small children are unlikely to be exposed to any appreciable degree. Also, if the leak were to be fixed, workers could directly contact the contaminated soil during excavation and removal of the soil. Similarly, if chemicals were present in groundwater at a site used for an office building, we could expect that office workers in the building would be the primary exposed receptors.

It is important to understand that identification of receptors does not specifically include any single individual, because the assumptions used to estimate exposure are used to describe a group or population of people, not an individual. It is very unlikely that any single individual will have exactly the same body weight, breathing rate, length of time spent at the site, etc. as those used to describe the population. Also, we are often interested in what could happen in the future. For example, using the gas station example, there is a chance the gas station will close and the site will be used in the future for a restaurant. We do not know who will be working and eating at the restaurant. Yet we know someone will, so we develop exposure assumptions that should be consistent with the types of people that could be there in the future (e.g., restaurant workers). Because of this lack of knowledge, receptors are typically referred to as "hypothetical" rather than actual people, meaning that exposures similar to these could occur, but that the exposure does not necessarily reflect any single person in that population.

Identify Complete Exposure Pathways

Once potential receptors are identified, the ways in which they may contact chemicals need to be defined. This is done by identifying complete exposure pathways. There are four components that must all be present for an exposure pathway to be complete:

- A source and mechanism of chemical release to the environment (e.g., leaking underground storage tank)
- An environmental transport medium for the chemical (e.g., movement of the chemical into air)
- A point of contact between a receptor and the chemical (e.g., indoor air in the attendant's station)
- An exposure route at the point of contact (e.g., inhalation)

Using the gas station example, let's assume that the chemicals are present in subsurface soil near the underground storage tank, and that the chemicals have migrated to groundwater. The groundwater is not used at the gas station, but is used by a neighboring residence that has a well on the property. Although the soil is contaminated, there is no way in which people could directly contact the soil, which is at depth beneath concrete. Therefore, while soil ingestion and skin contact with soil are potential exposure pathways, they are not complete (the third component of a complete pathway is missing). However, chemicals in soil could volatilize into air, where they could be inhaled by both employees and people visiting the station. Therefore, inhalation of chemicals from soil represents a potentially complete pathway of exposure. The same is true for groundwater on the site. Only inhalation of chemicals volatilizing from groundwater would be complete because the groundwater is not directly used. However, residents at the neighboring property might drink the water because they have a well. Therefore, ingestion of groundwater is potentially complete for the offsite resident receptor.

All of this information is generally depicted in a schematic diagram illustrating the four components of complete exposure pathways at a site. This schematic diagram is referred to as a Conceptual Site Model. An example of a Conceptual Site Model is shown in Figure 8.3.

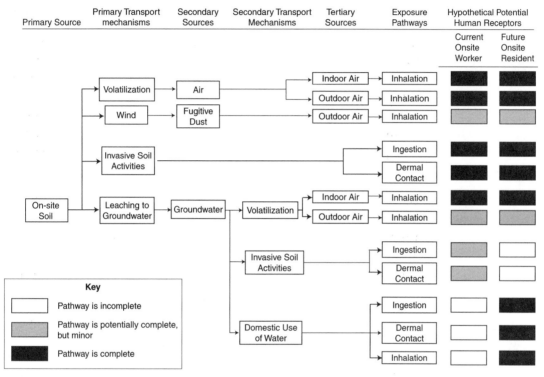

Figure 8.3. Sample Conceptual Site Model

Estimate Doses

Finally, a chemical-specific dose is estimated for each receptor and complete exposure pathway identified in the conceptual site model. In this step, assumptions are made regarding the rate at which exposure could occur. For example, it is typically assumed that the average American adult drinks 2 liters of water daily from the tap, weighs 70 kg, and incidentally ingests 100 mg of soil each day. It is further assumed that a resident lives at the same house for 30 years, and a worker works at the same location for 25 years. These assumptions can change if there is more specific information available about the site. Using the gas station example, the owner of the station could be contacted to obtain records about the job duration for a typical employee, and the number of hours a typical employee is present at the station each day.

The chemical concentrations identified in the data evaluation component are combined with all of these exposure or intake assumptions into equations to estimate a dose in units of milligrams chemical per kilogram body weight per day. These are the same units in which toxicity data are reported, as discussed in Chapter 7. For the gas station example, the results of this step would include chemical-specific dose estimates for all complete exposure pathways for each receptor. These dose estimates are combined with the toxicity values discussed below in the risk characterization component of a risk assessment.

Toxicity Assessment

In this component, toxicity values are identified for the chemicals at the site. Unlike the exposure assessment, toxicity values are the same regardless of the situation because toxicity can be defined only by the potency of the chemical. It is the combination of toxicity and exposure that determines risk and the likelihood of toxic effects. The methods used to develop toxicity values for chemicals were discussed in Chapter 7.

Toxicity values are separately developed for different exposure routes. Typically, values are developed for the oral and inhalation routes of exposure because the majority of toxicity studies are based on these exposure routes. In most cases, the oral values are used for the skin exposure route, adjusted for decreased absorption across the skin relative to the oral route. Two types of toxicity values are currently used in risk assessment: those describing cancer potency and those describing non-cancer effects. Some chemicals are known to have both cancer and non-cancer effects. In these situations, both cancer and non-cancer values might be developed. Therefore, each chemical might have up to four toxicity values. Toxicity values for cancer and non-cancer effects are discussed below.

Cancer Effects of Chemicals

A slope factor (SF) is used to quantify the cancer potency of a chemical. A SF defines the rate at which effects increase with dose. The higher the SF, the more potent the cancer effects of the chemical. An SF assumes that there is no dose that is associated with no risk. Any amount of the chemical could cause cancer, but the chance of this occurring might be so small as to be unmeasurable. This assumes that there is no threshold dose below which cancer does not occur (Chapter 7). We know this is true for some chemicals, and is likely not true for others. In general, this assumption ensures that we will not underestimate the potency of a potentially carcinogenic chemical to humans. These SF values are typically based on a conservative estimate of the cancer potency in test animals, and therefore include a safety margin in their derivation (see Chapter 7).

Dioxin has the highest slope factor of all chemicals regulated by the U.S. EPA, even though it is considered a probable, not a known, human carcinogen. Chloroform is one of the chemicals with a low SF. This SF is about seven orders of magnitude lower than that for dioxin (e.g., a factor of ten million), indicating that to get the same effect for both chemicals, someone must be exposed to 10 million times more chloroform then dioxin. Put another way, if the same concentration of both chemicals were present in the environment and the amount of exposure to both chemicals was the same, the cancer risk associated with the dioxin would be 10 million times higher than the cancer risk of chloroform. This demonstrates that we should not treat all chemicals equally because some are clearly more toxic than others.

Non-Cancer Effects of Chemicals

For the non-cancer effects of chemicals, a reference dose (RfD) is used. An RfD is a threshold dose below which adverse effects are not expected. The RfD is based on the assumption that adverse effects will only occur when certain physiological mechanisms are overcome, and will only occur above a certain level. Therefore, there are some doses that are indeed "safe" because toxic effects should not occur below these levels. In addition, we saw in Chapter 7 the numerous uncertainty factors used to generate these values, which has the effect of assuming the chemicals are substantially more toxic to humans than other species.

Unlike for SFs, the lower the RfD the more potent the chemical. One of the lowest RfDs is for tetraethyl lead, an organic form of lead that can enter the central nervous system and impact intelligence. This form of lead may have caused the mental problems of some of the early Roman emperors (see Chapter 1). One of the highest RfDs is for xylenes, a common ingredient of gasoline. RfDs for these two chemicals differ by a factor of about 10^5, or 100,000. This means that tetraethyl lead is 100,000 more potent than are xylenes because effects can occur from a tetraethyl lead dose 100,000 lower than a dose causing an effect from exposure to xylenes.

These basic differences between the toxicity values describing cancer and non-cancer effects are important because they impact the way in which the next step, risk characterization, is conducted. In addition, these differences make interpreting the results of the risk characterization more complicated.

Risk Characterization

This is the component of a risk assessment where risks are estimated by combining exposure and toxicity information for the chemicals at a site. There is an important distinction between cancer and non-cancer chemicals in this step. These two types of effects are discussed below.

Cancer Effects of Chemicals

To estimate a cancer risk, the dose estimated in the exposure step is multiplied by the slope factor identified in the toxicity step. The product of these two numbers is the cancer risk. This risk is defined as the probability of an individual developing cancer from exposure to the chemical, under the conditions assumed in the exposure assessment. This probability is considered an excess cancer risk, because the incidence of cancer from all sources other than the chemical is substantial. The annual incidence rate of developing some form of cancer is 4.41 per thousand people, or 0.00441 per person. This is a probability of 0.441 percent annually. Assuming the average person lives for 75 years and has this same probability of developing cancer each year, the probability of developing cancer over a lifetime would be 0.441 percent times 75 years, or 33 percent (Chapter 4). This means that each individual, on average, has a one in three chance of developing cancer over the course of his or her lifetime. This is from all sources that represent background conditions. For example, ultraviolet radiation from the sun is a potential source of cancer. There are naturally occurring levels of elements, such as arsenic, nickel, cadmium, or chromium, that may present a risk for developing cancer. Because the incidence rate for cancer is so high, the extra risk that chemical exposure might create is very rigorously managed.

Typically, sites in the United States are managed so that the extra cancer risk from chemical exposure is no greater than 1 chance in one million, or 1×10^{-6}. This means that, if one million people were exposed to the same concentrations as found at a site and all received identical doses over the same time period, one person would develop cancer due solely to the chemicals present at the site. In reality, the affected populations are typically much smaller than this and no cancers are expected to develop from chemical exposure, even for someone assumed to be highly exposed relative to others. To put this into perspective, if the background risk of developing cancer is 1 in 3 (0.33), then cancer risk from chemical exposure is managed so that the total risk of developing cancer is no greater than 0.330001. This is an unmeasurable increase in cancer risk because of the very small numbers of individuals involved at any one site. We will discuss the interpretations and significance of these numbers further in Chapter 9.

Non-Cancer Effects of Chemicals

For the non-cancer effects of chemicals, risks or probabilities of effects are not estimated. This is because, unlike the assumption of no threshold used for cancer chemicals, there are doses below which no adverse effects are expected from chemical exposure for non-cancer effects of chemicals. Instead of generating a risk, the estimated dose is divided by the reference dose, which is considered a "safe" level below which toxicity is not expected to occur. The result of this division is a ratio that will either be above or below one. This ratio is referred to either as a hazard quotient, if only one chemical is considered, or a hazard index if multiple chemicals are included. If the ratio is below one, this means that the estimated dose is below the threshold dose, and no toxicity is expected. If the ratio is above one, the estimated dose is above the threshold dose, and toxicity may result. Ratios above one do not mean that toxicity will result, but that there is a chance of this occurring. The higher the ratio above one, the greater the chance that toxicity would result from exposure.

Therefore, calculation of "excess risk" is only relevant for cancer effects of chemicals. For all other effects, there is either no risk (if the ratio is below one), or some degree of likelihood that adverse effects can occur (if the ratio is above one). This degree of likelihood for non-cancer effects of chemicals is typically referred to as "hazard." Regulatory agencies typically manage the non-cancer effects of chemicals so that concentrations at a site result in ratios no greater than one. Therefore, non-cancer effects would not be expected from chemical exposure at properly managed sites.

Given the conservative nature of the toxicity values developed for use in risk assessments (see Chapter 7), even hazards above one do not necessarily indicate that toxicity will result. However, because the goal of risk assessment is to protect all possible receptors, this target of one ensures that even sensitive members of a population (e.g., infants) are safe with regard to chemical exposure.

Ecological Risk Assessment

A similar process is followed for ecological risk assessment. Because of the tremendous number of different species and their different sensitivities to chemicals, the ecological risk assessment process is typically more complex and involves higher levels of uncertainty than human health risk assessments. As we discussed in Chapter 3, different levels of ecological organization (e.g., individual, community) might be exposed to and adversely impacted by chemicals. Less information regarding the toxicity of chemicals and exposure patterns is available for the great majority of species found in the environment than for humans. Only one species is targeted in a human health risk assessment; chemical-specific concentrations can be established for humans that would be below levels of concern based on the risk characterization process discussed earlier. In an ecological risk assessment, the receptors could include a mountain lion, deer, peregrine falcon, and wild oat plants. This only considers terrestrial species. If surface water is impacted by chemicals, then receptors could include fish, invertebrates found in sediment (e.g., crayfish), water fleas, and aquatic plants. Given the great diversity of species, the ecological risk assessment process is by necessity more of a framework than a cookbook. There are three primary components to the framework developed by the U.S. EPA:

- Problem formulation
- Analysis phase
- Risk characterization

In problem formulation, we identify the species that are relevant for a given situation, and identify the chemicals present in the environment. This component includes the data evaluation step discussed for human health risk assessment, but in addition, the ecological environment needs to be evaluated to identify the key targets of chemical exposure. The analysis phase includes the exposure and toxicity assessment components of a human health risk assessment, but again differs in several ways depending on the species involved and the ways in which exposure can be measured. Finally, in risk characterization either a likelihood of toxic effects is estimated, or a hazard quotient or index is generated. The framework for ecological risk assessment, which emphasizes the interrelationships among the components, is presented in Figure 8.4. Each of these components is further discussed below.

Problem Formulation

The problems and objectives related to an ecological risk assessment are more varied than for a human health risk assessment. At the beginning of a human health risk assessment, we know that the primary objective is to identify if cancer risk or non-cancer adverse health effects may result from chemical exposure above regulatory-based target levels. This objective is based on protecting the most sensitive individual that might be exposed. Things are not as clear-cut in ecological risk assessments. As we discussed above, there are a myriad of potentially exposed species at any one site, and these species differ in their sensitivity to chemicals and their overall role in the ecosystem. Therefore, the primary focus of the problem formulation step is to identify objectives of the ecological risk assessment. This includes the following steps:

- Integrate available information
- Select management goals and assessment endpoints
- Develop conceptual model
- Select measures of effect

Once these steps have been completed, the final step of problem formulation is to develop an analysis plan, which defines how the goals, endpoints, and measures of effect will be evaluated at the site.

To illustrate these steps, let's compare two sites with identical contamination in surface water and sediment. At one site, the surface water and sediment is part of a recreational fishing area (e.g., a lake stocked with trout). At the other site, the surface water and sediment are not used for recreational purposes, but osprey, protected raptorial birds, hunt fish in the impacted mountain lake. Different management goals, assessment endpoints, and measures of effect would be needed for these two sites, even though the type and extent of contamination are the same. In the first example, our goal would most likely be to protect the trout population in the lake. The ecological value of the lake community is primarily associated with the sport recreational use by humans. Therefore, we would manage this site to protect this resource. Individual trout might be affected, but this would be acceptable as long as the entire population is not affected. Variability in abundance is an acceptable measure of effect given the goal for the site.

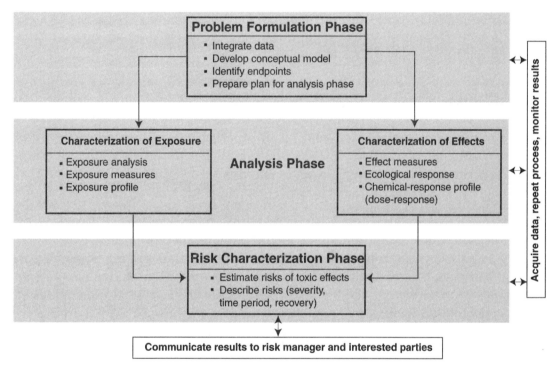

Figure 8.4. Overview of the Ecological Risk Assessement Process
[Adapted from U.S. EPA, 1998]

For the second example, the goal would be to protect individual ospreys from eating fish contaminated by chemicals in the surface water and sediments. Because the osprey is a protected species, we need to manage effects at the level of the individual in a risk assessment. Because the management goals and our endpoint species differ between the two sites, our assessment endpoints and measures of effect will also differ. Based on our perceived relative value of the two habitats, we can tolerate more effects on individual trout at the first site than we can to individual ospreys at the second site.

Identifying what is of value and what needs to be protected is a critical, complex, and sometimes controversial process in an ecological risk assessment. The ecological risk assessment process includes involvement of trustees, which are agencies or individuals that have a regulatory or ownership issue associated with the site. For example, the United States Fish and Wildlife Service would have a regulatory issue at both sites because protection of wildlife is involved. The Department of Parks and Recreation would have regulatory involvement for the lake that is fished. If one of the sites was on Indian land, the Bureau of Indian Affairs would also have a stake in the process. Each of these stakeholders could provide input into the selection of management goals, assessment endpoints, and measures of effect that define the scope of an individual project. This is the primary difference between human health and ecological risk assessments. In a human health risk assessment, the goals and objectives of the assessment are generally known before the process starts. In an ecological risk assessment, they are never pre-defined. Instead, they are developed separately for each site.

Analysis Phase

There are two components to the analysis phase: characterization of exposure and characterization of effects. Overall, this phase is similar to the exposure and toxicity assessment components of a human health risk assessment. The primary differences relate to the variety of ways in which exposure and toxicity can be measured in an ecological risk assessment.

Characterization of exposure includes identifying measures of exposure and conducting an exposure analysis, and culminates in development of an exposure profile. Exposure can be based on assumed intake assumptions, as is done for human health risk assessments, or can be based on direct measurements from animals or plants collected at the site.

For example, assume you want to estimate chemical exposure by a meadow vole, which serves as the food source for many predators. The goal of collecting this information is to set "safe" concentration levels in the soil to protect the predators. Using the human health approach, intake assumptions would be identified for the amount of food and soil consumed by the predators daily, and concentrations in food would be estimated from soil concentrations using models. If we assume that the predators are foxes, we would need to estimate chemical concentrations in meadow voles because this species is a staple in the diet of some foxes. This approach requires a substantial number of assumptions and use of many models, each of which introduces uncertainty in the results. Alternatively, meadow voles could be collected and chemically analyzed to directly measure chemical concentrations. These measured concentrations could then be used to estimate exposure by foxes, eliminating some of the uncertainty in the estimate.

Characterization of effects includes identifying measures of effect and conducting an ecological response analysis, and culminates in development of a chemical-response profile. Toxicity can be based on literature values (but usually only available for different species), or on bioassays conducted in the laboratory on soil, sediment, or water from the site.

For example, assume that a pond is contaminated with metals. One goal of the ecological risk assessment is to identify if a known sensitive aquatic species (e.g., freshwater shrimp) is impacted by the concentrations of metals in the pond. This shrimp is a source of food for predatory fish in the pond, and therefore its abundance is linked to the health of the community. Literature values could be used to identify "safe" concentrations of metals for various species, but not for the specific one of interest at our site. These could be used and extrapolated to the target species, which introduces uncertainty into the results. Alternatively, water and/or sediment from the site could be brought into the laboratory and the species of interest could be directly tested for toxicity. These studies are known as bioassays. Results of bioassays are used to develop a protective concentration relevant to the species and site of interest.

Both of these processes are iterative in nature. In addition to the individual-level exposure and toxicity assessments conducted in human health risk assessments, an ecological risk assessment also can include evaluation of community-level receptor exposure and toxicity, if appropriate based on the Problem Formulation step.

This process allows for more flexibility than is available for human health risk assessments. Depending on the goals of the ecological risk assessment, the methods used to measure exposure and toxicity may differ.

Risk Characterization

The risk characterization step involves two components: risk estimation and risk description. The risk estimation component is similar to the human health risk characterization conducted for non-cancer effects of chemicals in that it quantifies potential effects from chemical exposure. Depending on the methods used to estimate exposure and toxicity, the methods used in risk estimation for ecological receptors may differ from those used for humans. One method that can be used, which is similar to the method used for humans, is the toxicity quotient method. In this method, the estimated exposure is divided by a "safe" level of exposure developed in the characterization of effects component. The resulting value is compared to a threshold level of one. Below this level, no effects are expected (regardless of what the impact might be). Above this level, there may be effects.

Another added feature in ecological risk assessments is the risk description component of risk characterization. In this component, nature and intensity of effects, their spatial and temporal scale, and the potential for ecosystem recovery are all addressed. This component partially serves to identify ways to remedy effects at a site.

Returning to the above examples, assume that the results from the first site (the lake stocked with trout) indicate that the trout population is declining, and that this has been occurring over the past 3 years. However, the rate of population loss has gone down each year, indicating that the impact has been declining over time. This is expected from contamination where the chemicals degrade over time. Under this scenario, the remedy with the least overall impact to the ecosystem might be no action. The impact to the ecosystem of actively removing contaminated sediment from the lake would be greater than that if the chemicals were allowed to degrade on their own. This can be due to several factors, including the following:

- Physically moving the sediment will liberate some of the chemicals into the lake, increasing the availability of the chemicals
- There would be physical effects on the habitat due to use of heavy machinery to dredge the sediment and transport it away from the site
- Species residing in the sediment will be removed from the lake; this could impact the lake community for several years until population levels in the sediments recover

For the second site, let's assume that the osprey might be impacted because the DDT in the sediments impacts their reproductive ability. The lake is a prime breeding ground for ospreys. Under this scenario, a no-action decision would only hasten the decline of the birds because DDT degrades very slowly in the environment. Instead, at this site, active removal of the DDT-contaminated sediment would be preferential. Although a season of breeding might be interrupted, the problem would be eliminated in the future. This would result in a lower overall impact to the birds.

9

Communicating Risks to the Public

Risk communication is the process of making risk assessment and risk management information understandable to the layperson, including community groups, environmentalists, and lawyers. These groups and professions typically want to know if something is safe, not how uncertain the risk may be. As we discussed in Chapter 8, results of risk assessments are often highly uncertain due to lack of knowledge. In part, the purpose of this book is risk communication. We have discussed many of the key components of toxicology, and provided many examples to illustrate these components. In spite of this knowledge, decisions about whether or not something is safe are rarely based purely on science. Instead, our ideas about chemical toxicity and risks from exposure are based on our concept of risk perception.

People and communities respond differently to potentially hazardous situations. An event considered acceptable by one person might be unacceptable to another. This is a behavioral response. Understanding these behavioral responses is key in developing risk management decisions. Risk management is the process by which policy is set based on results of risk assessments or toxicity testing. Risk management decisions typically take into account the scientific information, but often are primarily based on public perception. For example, lead can be placed underneath future home sites and exposure to lead after the homes are built would be essentially zero because there is no way for the residents to contact the chemical in soil beneath the home. If the concentration of lead under the home were 50 parts per million (essentially background levels), few people would be concerned. However, if the concentration of lead were 10,000 parts per million, most people would be concerned. The risks in both cases are essentially identical and below any levels of concern, but the public perception would prevent the second option from being implemented.

This issue of public perception has been the focus of numerous studies, including studies by the League of Women Voters in 1979, the National Research Council, and individual scientists and psychologists. My colleagues and I have also conducted similar studies of undergraduates, graduate students in science and other disciplines, and coworkers. Some key points of these studies are summarized below to provide a basis for the rest of this chapter.

In the League of Women Voters' study, the general members, active league members, students, and scientific experts were asked to rank 30 activities or agents in order of their risk of dying. The results were very different for these different groups:

- Laypeople ranked motorcycles and handguns as most risky and home appliances, power mowers, and football as least risky
- Active league members ranked pesticides, spray cans, and nuclear power as safer than did the laypeople
- Students ranked contraceptives and food additives as risky and mountain climbing as less risky than did other groups
- The scientific experts ranked nuclear power and police work as less risky than did other groups, but considered electric power, surgery, swimming, and x-rays as more risky than did the other groups

In unofficial studies conducted by myself and colleagues, the public ranked stratospheric ozone depletion, worker exposure to chemicals, hazardous waste, industrial pollution accidents, and oil spills as of high concern. However, experts only agreed with the public rankings for ozone depletion and worker exposure to chemicals. While the public considered indoor air pollution (including radon) of low concern, experts ranked this as of high concern.

These differing perceptions impact what risk management decisions are acceptable in different situations. The perceptions relate to psychological factors such as fear, dread, uncontrollability, and voluntary versus involuntary risks, as well as familiarity with the hazard. Figure 9.1 illustrates the relationship of these factors for various hazards. Hazards in the upper right quadrant are those that are perceived as both relatively unknown and uncontrollable, and include asbestos, mercury, pesticides, and radioactive waste. The hazards in this quadrant are most likely to be considered most "dreadful," and are most likely to lead to government regulation based on public concern.

We discuss this "risk space" in more detail below.

Voluntary and Involuntary Risks

A key component of risk perception is whether or not the risk is considered voluntary. Absolute risk is less important to an individual than whether or not the risk is voluntarily chosen. For example, some relative risks of dying are shown in Table 9.1 for voluntary activities. Relative risks of dying from various involuntary activities are shown in Table 9.2. The highest average annual risk of developing cancer from

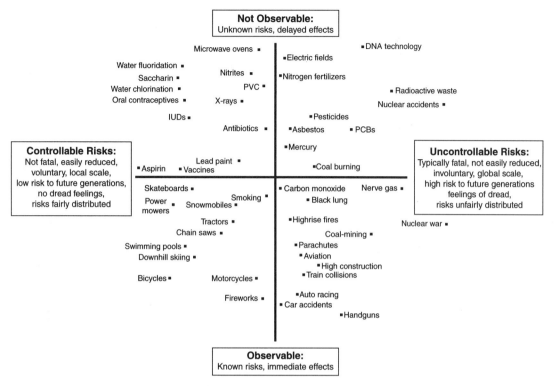

Figure 9.1. Risk Space Diagram
[Modified from Klaassen, 1996]

any one of the involuntary sources (e.g., cosmic rays while on an airplane) is about 5 in 1,000,000 (5×10^{-6}). Risk of dying from voluntary activities are typically much greater than these risks of developing cancer from involuntary sources. For example, the annual death rates for motorcycling, horse racing, and smoking are all greater than 1 in 1,000 (1×10^{-3}). Nevertheless, these involuntary risks are not typically accepted while the higher voluntary risks are accepted. It is human nature to accept some risk if one chooses; it is the lack of choice that is not tolerated.

In risk management decisions for chemicals based on risk assessment, a target cancer risk of 1 in 1,000,000 (1×10^{-6}) is typically used for chemical exposure. This is an involuntary risk and therefore is not as well tolerated as voluntary risks. This level of risk was established by regulatory agencies as a level below which was considered a "trifle" and not of regulatory concern. This is approximately the same risk of being struck by lightning. U.S. EPA considers this 1 in 1,000,000 risk as "the maximum lifetime risk that is essentially zero."

Table 9.1. Relative Risks of Dying from Voluntary Activities

Activity	Risk of Dying (probability per person per year)
Motorcycling	1 in 50 (2×10^{-2})
Horse racing	1 in 740 (1×10^{-3})
Cigarette Smoking[a]	1 in 200 (5×10^{-3})
Power boating	1 in 5,900 (2×10^{-4})
Rock climbing	1 in 7,150 (1×10^{-4})
Auto racing	1 in 10,000 (1×10^{-4})
Consuming alcohol[b]	1 in 13,300 (8×10^{-5})
Professional boxing	1 in 14,300 (7×10^{-5})
Use of birth control pills[c]	1 in 50,000 (2×10^{-5})
Canoeing	1 in 100,000 (1×10^{-5})
Skiing	1 in 1,430,000 (7×10^{-7})

[a] Twenty cigarettes daily.
[b] One bottle of wine daily.
[c] From adverse acute reaction to medicine.

Table 9.2. Relative Risks of Dying from Involuntary Activities

Activity/Source	Risk of Dying (probability per person per year)
Second-hand smoke at home[a]	5 in 1,000,000 (5×10^{-6})
Eating peanut butter[b]	4 in 1,000,000 (4×10^{-6})
Being struck by lightning	1 in 1,000,000 (1×10^{-6})
Cosmic rays on airplane[c]	1 in 2,000,000 (5×10^{-7})

[a] Sharing a room with someone who smokes.
[b] Four tablespoons daily; lung cancer due to aflatoxin, a natural toxin produced by a bacteria (*Aspergillus flavus*) that can infect stored peanuts.
[c] From cancer developed as a result of exposure to cosmic rays.

Approaches to Risk Communication

There are several general statements that can be made regarding the general perception of risk. These are summarized as follows:

- Voluntary risks are accepted more readily than those that are imposed
- Risks under individual control are accepted more readily that those under government control
- Risks that seem fair are more acceptable than those that seem unfair
- Risk information that comes from trustworthy sources is more readily believed than information from untrustworthy sources
- Risk that seems ethically objectionable will seem more risky than those that do not
- Natural risks seem more acceptable than artificial risks
- Exotic risks seem more risky than familiar risks

We have already discussed the first of these points. The second point relates to the controllability of the risk. As shown on Figure 9.1, risk associated with over-the-counter drugs like aspirin, antibiotics, and oral contraceptives will be more accepted than those associated with air pollution, pesticides, or electric fields because we have more control over the former. Even if the risks are identical, the ability to control our exposure to some degree makes the risk seem lower. This is one reason why reports like those discussed in Chapter 2 for Alar, a plant growth regulator formerly used on apples, lead to concern over possible risks. The use of Alar on apples was allowed under United States regulations, which is considered less tolerable than if the decision were made by the general public to allow its use on apples.

The third point listed above, that risks which seem fair are more acceptable than those that seem unfair, is less tangible to identify. One example from Figure 9.1 is related to electric fields. Although the risk from electromagnetic radiation associated with overhead power lines might be low, the decision to allow their continued, uncontrolled use can be considered unfair to those living near such lines. Because the risk is not equal for all people, it can be considered unfair to some. Such risks are less acceptable than those that equally affect everyone. Another example could be chemical cleanup levels at Superfund sites. Although the risk of toxicity from chemical exposure might be low, people living near the sites could consider any solution that leaves chemicals in soil at such sites to be unfair because of their proximity to the site. The proximity to the site can increase the perceived risk, which leads to concern over the solution.

The fourth point listed above is that risk information that comes from trustworthy sources is more readily believed than information from untrustworthy sources. This is often overlooked, but can have important consequences to risk management decisions. Reports have been published that indicate various disciplines and authority figures are associated with different levels of trust. These are summarized below as they relate to environmental issues:

- High level of trust:
  - Professors and researchers
  - Doctors
  - Clergy
  - Company employees

- Moderate level of trust:
  - Media
  - Environmentalists
  - State governments (generally)

- Low level of trust:
  - Federal government
  - Industry (as a whole)
  - Consultants to government and industry

Considering that the majority of information related to toxicology and risk we receive comes from the media and government, there is a built-in degree of mistrust in many of the things we hear and read. The categories listed under high degree of trust might not be the people with the most knowledge or complete understanding of a given situation. However, we still rely on them more than others due to the trust factor. This will be further discussed below.

The next point is that risks that seem ethically objectionable will seem more risky than those that do not. An example of this is the lead in soil beneath homes discussed earlier in the chapter. It can be considered unethical to leave toxic chemicals beneath somebody's home. Even if the risk is essentially zero, it nevertheless is not an accepted solution because of ethical considerations. Another example is related to asbestos abatement in buildings. Asbestos found in occupied buildings is removed and disposed of, even though the act of removing the asbestos can increase the risk of toxicity over leaving it in place in the building. But leaving it in place, especially in schools, can be considered unethical because a known lung carcinogen is in a place visited by children. The fact that the chances of exposure are negligible is beside the point to those bothered by the ethics of the situation.

The next point is that natural risks seem more acceptable than artificial risks. Natural risks include lightning, tornadoes, earthquakes, cancer from cosmic radiation, and ingestion of natural products that are toxic (e.g., puffer fish, deathcap mushrooms). Artificial risks are those associated with manufactured chemicals or situations, such as pesticides in soil or air pollution from industrial emissions. Even if the risks are identical, the artificial one will be considered less acceptable than the natural one.

The final point listed above is that exotic risks are considered more risky than familiar risks. This is shown in Figure 9.1, where exotic risks are on the bottom of the risk space and familiar risks are above the horizontal line. Risks that are more familiar are more acceptable because we have grown accustomed to their presence. This is in part fueled by the fear of the unknown. We will generally be more afraid of things we do not understand than things we do understand. While we drive automobiles and understand the risks associated with this, some tribes in Africa might have never seen an automobile. The reaction of these natives is likely to be one of fear related to the noise, smell, and speed of the vehicle. While we might laugh at such a response, it captures the differences between acceptability of risk between familiar and unfamiliar events.

In addition to these points, three other factors have a predictable impact on public perception of risk. These are:

- Risks are more acceptable when the benefits are recognized and clear
- Risks are more acceptable if there are no alternatives to reduce risk
- Risks are less acceptable if they involve primarily children rather than everybody

The first factor is one reason why most pesticides used in agriculture are accepted, even though the chemicals used have some degree of risk associated with their use. Without pesticides, the crop production we have enjoyed over the past decades would have been reduced due to losses from insects and weeds and food prices would increase. Because the benefits of pesticide use have been clearly identified, the risk associated with their use is usually tolerated. Risk associated with other chemicals for which benefits are not clearly defined, such as MTBE, are less accepted. If it can be clearly demonstrated that use of MTBE in gasoline reduces the levels of benzene, a known human carcinogen, in air, then its use might be more acceptable (see Chapter 2 for more discussion of MTBE).

The second factor is straightforward. Because risk management decisions involve selecting one of several alternatives, selecting an alternative that leaves chemicals in soil and/or groundwater is less acceptable, from a public perception standpoint, than completely removing the chemicals. Even if the costs of removing the chemicals are prohibitive, this choice is preferred by the public over leaving chemicals in place.

Finally, risk that primarily focuses on children is always less acceptable than risk that equally affects everyone. A good example of this is lead. Children are most sensitive to the toxic effects of lead, and these effects can lead to impaired intelligence later in life. Because of this, lead toxicity is a very emotional issue for many people.

Considering all of these factors and issues, different groups have different approaches to communicating risks to the public. Three of the most common are the media, regulatory agencies, and contractors and consultants. The latter of these groups typically represent industry. As listed above, the media and state agencies have a moderate degree of trust with the public, while contractors and consultants for industry have low trust. This also impacts how the different groups approach risk communication. Understanding these differences helps in interpreting the information the different groups present to the public. The following discussion presents generalities; there will be exceptions within each group.

Media

The primary goal of the media in risk communication is alerting the public to a threat. Rarely does lack of toxicity capture the public's attention. Instead, cause for concern might capture listeners or readers. Therefore, the media will tend to focus on the pessimistic or "worst-case" evaluation of a risk issue. Using the factors discussed earlier in this chapter, the media will tend to focus on the following:

- Involuntary risks (i.e., imposed risks)
- Risks under government control
- Risks that seem unfair
- Risks that seem ethically objectionable
- Artificial risks
- Exotic risks

MTBE was discussed in Chapter 2 as an example of inaccurate and incomplete media reporting. Let's look at this example in more detail as it relates to these factors. MTBE was put into fuel by oil companies based on a requirement for "fuel oxygenates" by the federal government. Therefore, the first two factors listed above are met for MTBE coverage by media. MTBE effects that have made the news have generally been related to either (1) groundwater in areas where it is used as a drinking supply, or (2) surface water in reservoirs and contamination from jet skis. Coverage of these two types of MTBE contamination make the point that the risks of MTBE are unfair because it only affects those using the water. This makes MTBE contamination an "us versus them" issue, which makes a good story and feeds the public idea of unacceptable risk. Because MTBE smells and tastes bad at levels far below those that cause toxicity, the media can use the point that it is ethically objectionable to allow MTBE use when it has an aesthetic impact. Because the chemical is manufactured and used by industry, risks associated with its use are considered artificial and thus are considered less acceptable than risks from natural chemicals. Finally, the public's lack of knowledge about this new fuel additive adds to the fear associated with exotic risks relative to common risks.

Trust also comes into play. In several media reports, MTBE was presented inaccurately as a human carcinogen. The public will generally believe the media report unless a figure with a higher trust level, such as a university professor, contradicts it. If industry personnel or consultants present the public with factual information regarding MTBE's lack of carcinogenic activity, public opinion is unlikely to be swayed because this group has a lower trust level with the public.

The preceding discussion is not meant to decrease the importance or relevance of the information presented by the media; it is meant to allow the public to better understand the basis of the information and realize that the media may at times be presenting only part of the story, and perhaps only those issues that will get the attention of the viewer or reader. While most journalists want to get to the truth and report information in the least biased way possible, they are people and they have the same fears of the unknown as do the rest of us. Exotic risks, for example, will seem more risky to them than common risks.

In addition, the media generally does not understand the risk assessment process, which we discussed in Chapter 8. As previously discussed, there are many problem areas in health risk assessment, many of which are listed below:

- Threshold versus non-threshold dose-response
- Animal to human extrapolation
- High-to-low dose extrapolation
- Extrapolating cell culture results to a whole organism
- Extrapolating laboratory data to the field
- Estimates of exposure
- Interactions across chemicals
- Toxicity of complex mixtures
- Definition of "acceptable" risk

The media is largely ignorant of these relevant issues. As a result, they are often unable to differentiate between real health issues and those of low risk. Alar, discussed in Chapter 2, is an example of the failure to differentiate between these two. The media is valuable in that it alerts the public to possible concerns. However, the public needs to be able to understand when something is only a possible concern and when it is a real concern.

Regulatory Agencies

Regulatory agency personnel have a very different focus in risk communication. The job of agency personnel is, in general, to enforce state and federal environmental regulations and protect human health and the environment. Chemical concentrations are either above or below levels allowable under the regulations. Agency

personnel must ultimately approve all chemical remediation at sites, approve decisions to require no further action in relation to chemicals at sites, and make risk
management decisions. The risk management decisions must be consistent with
applicable environmental regulations, and, because agency personnel are public
servants, should be acceptable to the public. This puts agency personnel in the
difficult role of supporting all risk management decisions they make, while still
allowing responsible parties such as industry to minimize costs associated with
cleanup of chemicals in the environment. Unlike the media, many agency personnel
understand the problem areas of risk assessment, and it is their job to communicate risks to the public.

Also, agencies often must deal with environmental groups that want all chemicals
removed from the environment, regardless of the cost or regulatory requirements.
Agencies may be sued by environmental groups if they consider a risk management
decision to be unfair or not sufficiently protective. Therefore, agency personnel are
often hesitant or reluctant to approve risk management decisions that allow for
chemicals to remain in the environment, even if risk assessments show that risks
are below regulatory levels of concern.

For example, a risk assessment was conducted on DDT remaining in soil 15 years
following its use for mosquito abatement in the 1970s. The consultant's and the
agency's toxicologists agreed that leaving 30 parts per million DDT in the soil was
adequately protective to meet regulatory requirements. At a public meeting, those
attending seemed agreeable with the decision. However, when the final risk management decision was made by the agency, a level of only 5 parts per million DDT in
soil was allowed to remain at the site. The explanation from the agency project
manager was that no levels higher than 5 parts per million had been allowed to
remain in soil at any other site in his jurisdiction. It is likely that this decision was in
part made to minimize the potential for environmental groups to sue the state over
the "high" level approved by the agency in the risk assessment.

Public perception is often the primary factor in agency selection of a risk management decision for chemical cleanup at sites.

Contractors and Consultants

Of the three groups discussed in this chapter, the level of distrust by the public is
highest for contractors and consultants. Because consultants typically work on
behalf of industry, the public assumes that the decisions they reach are all favor-
able to industry, and are designed to either deceive the public or downplay concerns over chemical exposure. Although this is true in some situations (e.g., tobacco company consultants saying that nicotine is not habit-forming), this is usually unfounded. Consultants must have their work reviewed and technically accepted
by agency personnel. This requires a certain level of objective scientific analysis.

As we have discussed throughout this book, there are many uncertainties related to toxicology and risk assessment. In most cases, regulations dictate that these uncertainties be dealt with in a conservative manner to ensure that harmful chemicals are not left where they can lead to toxic effects in people or impact the environment. Also, agencies typically do not have sufficient funds to conduct the extensive toxicology research that would reduce these uncertainties. Part of a consultant's job is to identify these uncertainties and minimize them as much as possible. Because the assumptions used in the risk assessment are conservative, reducing uncertainty tends to increase chemical concentrations that can be considered safe. Next, the consultant needs to convince the agency, based on technically sound science, that the concentrations are protective. This was the case for the DDT site discussed above. Because the proposed cleanup levels are often higher than the conservative levels developed by agencies, the public is inherently leery of the conclusions. Industry pays for research to reduce uncertainties because this is often more cost-effective than spending millions of dollars to clean up a site to background levels, or to the overly conservative numbers developed by agencies with limited funds.

As a result of these inherent fears of and bias against consultants, a formalized process has been developed to plan risk communication strategies for consultants and industry. The following questions are typically included when a strategy is developed:

- Who is your primary audience?
- What does your audience already know?
- What is the main message you are trying to communicate?
- What do you want your audience to learn?
- What are the points your audience will most likely misunderstand or get wrong?
- What sources of information are trusted by your audience?
- What does your audience need to know?
- What doesn't your audience know?
- What kinds of risks are involved?
- Are people likely to underestimate or overestimate their risk?
- Is this kind of risk likely to cause either outrage or apathy?
- What is the best way to get the message across (video, television, meeting, etc.)?
- How will success of the program be defined and measured?

An effective risk communication strategy will be one that answers these questions truthfully and accurately. The method in which information is communicated is at least as important as the information itself. For example, consider a simple written poll that asks people if they "favor or oppose new electrical transmission lines." For one such poll, 47 percent of respondents opposed the new lines while only 38 percent favored them (the other 15 percent were undecided). The same poll was conducted on a different sample of people from the same area, but was worded differently. In this second poll, people were asked if they "favor or oppose new electrical transmission lines to meet future growth of the area." In this poll, 54 percent of the respondents favored the lines while only 35 percent opposed the lines. The difference between the questions asked in the two polls was minor, but the second poll provided an explanation of why the new lines were needed. In this case, the benefits of the lines were recognized and clear. As discussed earlier, this is one factor that increases the acceptability of risk to the public.

Concluding Thoughts

The topic of risk communication is a highly emotional one for most people. Pure science has a role in risk communication, but compassion, the ability to listen and speak clearly, and gaining credibility are probably more important. In general, everyone associated with a situation involving potential risk of chemical exposure has the same goal. This goal is to ensure the health and safety of humans and protection of the environment. Although many factions may have different motivating factors (e.g., selling newspapers, limiting liability, saving money), the endpoint is often the same.

The public perception of risks associated with chemicals is largely based on lack of knowledge and on basic human psychology. Too often risk communication has either involved demeaning the general public or talking over their heads. These methods of communication only exacerbate the problems. The ability to communicate risks to the public in an understandable manner is one of the best ways toxicologists can demonstrate the importance of toxicology in our everyday lives.

10

New Approaches in Toxicology and Risk Assessment

Toxicological knowledge is gained daily. This new information is used to refine, improve, or modify our hypotheses about how chemicals act to cause toxic effects in humans, and at what levels of exposure. Because risk assessment techniques are based in part on toxicological information (e.g., dose-response curves), some of the changes to these hypotheses can impact how risk assessments are conducted. Risk assessment is evergreen because its methods change as the science behind the methods evolves. Many changes are minor, and might involve things like modifying the breathing rate assumed for an active adult male or the toxicity value of a chemical by a small amount (e.g., a factor a two). Other changes can be more significant. Significant changes occur to risk assessment guidelines when enough new toxicological data have been collected to demonstrate that the current method is either no longer relevant, is scientifically inaccurate, or is outdated.

This chapter presents four examples of significant developments in risk assessment primarily brought about by new toxicological data collected from animal studies. The first two provide alternate methods for identifying cancer and non-cancer toxicity values, respectively, for use in human health risk assessments. In both cases, new information indicated that different methods were needed to reduce the uncertainty surrounding the toxicity value, and that current methods were based on outdated science. The third example provides an alternative method for quantifying risks in a risk assessment, and the final example considers a newly discovered area of toxicology and its potential impact on future toxicology and risk assessment methods.

This chapter is not intended to present an overview of all recent developments in toxicology, but through the use of examples attempts to illustrate the breadth and variety of toxicology research and applications that touch our everyday lives.

Use of Thresholds in Estimating Cancer Effects

A principal tenet of risk assessment developed in the 1980s is that any exposure to a chemical that is considered carcinogenic is assumed to present a risk of cancer, no matter how small the dose. That is, there is no threshold associated with cancer. When it was first made in the 1970s, this assumption seemed reasonable in order to protect human health because we had relatively little knowledge about how cancer worked at that time. Since then, much data have been accumulated that suggests some chemicals may not act as carcinogens below a certain dose level. This dose level is referred to as a threshold. For example, we know that thresholds exist for at least one of the four stages of carcinogenesis (promotion). This knowledge, combined with the fact that there are repair mechanisms within the body that attempt to reverse or minimize damage from chemical insult, implies that thresholds are likely to be associated with certain types of cancers for some chemicals.

Integrating this information into the risk assessment process would mean that a chemical could be treated as a carcinogen if the concentration were above the cancer threshold level, but could be treated as a non-cancer chemical at lower concentrations. Regulating a chemical in this way would allow for more accurate consideration of toxicology when setting a level that is considered "safe" to humans.

Arsenic and chloroform are two chemicals that might have thresholds for skin cancer through water ingestion and liver cancer, respectively. We will examine arsenic in more detail as an example of the impact this approach can have on identifying risks and setting "safe" levels for a chemical. None of the approaches discussed below have been incorporated into any U.S. EPA-based toxicity value at this time. However, new cancer risk assessment guidance recently released by the U.S. EPA allows for consideration of thresholds for cancer in establishing slope factors, which opens the door for more sophisticated and accurate methods to be integrated into the risk assessment paradigm.

In the past, due to lack of knowledge regarding cancer potency at low concentrations, a conservative model has been used to identify slope factors from high dose data in animals.

Probably the most common of these models is the Linearized Multistage (LMS) model. As discussed in Chapter 7, this model assumes the cancer potency of a chemical is linear at low doses, as shown in Figure 10.1. A slope factor developed using this model will usually lead to an overestimation of the true cancer potency of a chemical. The slope of the line in our example is shown as the dotted line in Figure 10.1. The 95UCL of the slope of this curve is then used by U.S. EPA as the slope factor to set "safe" concentration levels. For arsenic, this concentration is about 45 parts per billion in drinking water.

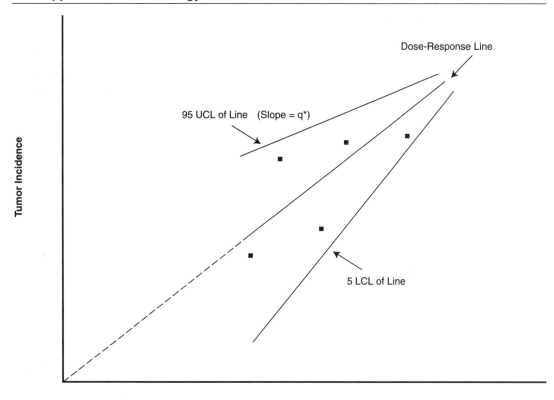

Figure 10.1. Use of the Linearized Multistage Model to Develop Cancer Potency for Use in Risk Assessment

Many epidemiological studies (see Chapter 5) have attempted to relate exposure to arsenic through drinking water to skin cancer. Studies conducted in Taiwan and Chile conclusively indicate a link between the two; this information has led to arsenic being classified as a known human carcinogen. However, similar results have not been seen in the United States. Results of several of these studies are shown in Figure 10.2. The studies from Asia and South America are all on the right half of the curve (e.g., had the highest arsenic concentrations in drinking water). All of the American studies had lower arsenic concentrations in drinking water (left half of the graph in Figure 10.2), and none of these saw a relationship between arsenic and skin cancer. However, the sample sizes of these American studies were very small compared with the others.

There is an obvious change in the relationship shown in Figure 10.2 at an arsenic drinking water concentration of about 0.1 mg/L, corresponding to a daily dose of about 0.004 mg/kg/day for an average adult. As previously discussed (Chapter 4), arsenic is metabolized to a nontoxic form by a specific enzyme. This enzyme requires a cofactor (another chemical produced by the body) to function. If exposure to arsenic is high enough, the cofactor can be depleted, which in turn prevents the enzyme from functioning. If the enzyme does not function, then arsenic is not detoxified and it can exert effects. The dose level at which this metabolic detoxification pathway becomes saturated is estimated to be 0.007 mg/kg-day, or 0.5 mg/day for an average adult. Using typical risk assessment assumptions, this corresponds to a water concentration of about 0.25 mg/L.

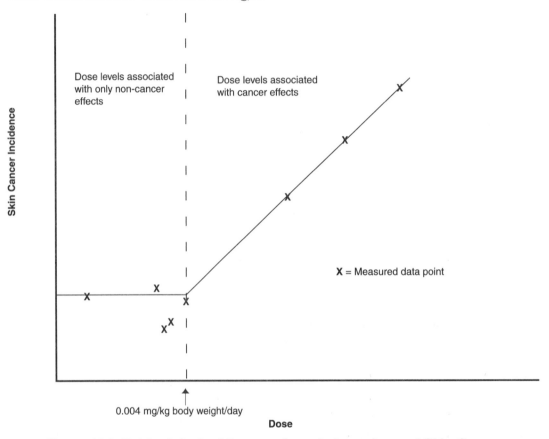

Figure 10.2. Epidemiological Data on Arsenic Ingestion and Skin Cancer

Based on this information, the dose level at which cancers first start to be seen from arsenic exposure is near the level at which the detoxification pathway is overwhelmed. Only epidemiological studies with arsenic concentrations in drinking water at or above 0.6 mg/L have reported incidence of skin cancers that increase with dose. The lowest estimated dose corresponding with positive studies of 0.011 mg/kg/day is slightly above the metabolic saturation threshold dose. This is solid evidence that arsenic does not act as a carcinogen in drinking water below this level.

The regulatory impact of such an approach is shown in Table 10.1. Under existing methods, the arsenic in drinking water concentration considered "safe" by U.S. EPA is 0.005 mg/L. This is the concentration associated with an acceptably low cancer risk, assuming there is no threshold for cancer. Using the threshold approach, a concentration of arsenic in drinking water of 0.23 mg/L would be considered "safe." In this case, the 0.23 mg/L concentration is the threshold for non-cancer effects. Although non-cancer effects can occur at levels below the threshold for cancer (repair mechanisms for liver damage might not be as effective as they are for DNA), this is not the case for arsenic. Below concentrations of 0.070 mg/L in drinking water, arsenic would be regulated only as a noncarcinogen. Based on the threshold concentration for effects discussed above, water concentrations below 0.070 mg/L would not require any cleanup. Concentrations above this level would be evaluated and regulated based on cancer risks.

**Table 10.1. Regulatory Impact of a Skin Cancer
Threshold for Arsenic in Drinking Water**

Assumption[a]	Target Drinking Water Concentration[b]	Basis for Target Level
No Threshold	5	Excess cancer risk = 1 in 10,000 (1×10^{-4})
Threshold	230[c]	Noncancer Hazard Quotient = 1

ug/L = micrograms of chemical per liter of water.

[a] Used to describe the dose-response relationship for cancer effects.
[b] Parts of chemical per billion parts of water (ug/L). Values assume ingestion of 2 liters of water daily for a 70 kg adult over a 30-year period.
[c] If the concentration is above 230 ug/L, treat arsenic as a carcinogen.

But even if convincing evidence were available that many chemicals act as carcinogens only above certain levels, it is difficult to set "safe" levels for regulatory purposes using such information. In part, this is due to the inherent variability in the human population. The threshold level shown for arsenic in Figure 10.2 might be higher or lower for any specific person.

Other types of cancer dose-relationships are also being identified for some chemicals. The EPA's new cancer risk assessment guidance is more flexible to allow use of these chemical-specific models in an effort to incorporate the best science into selection of "safe" concentrations.

Use of Benchmark Doses in Estimating Non-Cancer Effects

As we discussed in Chapter 7, toxicity factors used to evaluate non-cancer effects are primarily based on no-observed adverse effect levels (NOAELs) from animal studies. Uncertainty factors are incorporated into the NOAEL to develop a reference dose that represents a threshold dose below which adverse effects should not occur. This dose can be considered a "safe" dose. However, this approach suffers from several limitations, including the following:

- The maximum NOAEL (i.e., the threshold dose) is not usually identified; instead the NOAEL is identified based on a dose level chosen for the study
- The dose-response relationship of the chemical is ignored; only the NOAEL is used

By definition, the NOAEL is a level at which no adverse effects occur. However, all levels below the threshold will be NOAELs. The goal of animal studies is to identify the highest dose that causes no effect. This is the threshold dose, and is the most relevant dose upon which to base "safe" concentrations. Use of the NOAEL rarely allows for use of this dose; in most situations the investigator cannot know what the highest NOAEL is because it would require testing of too many different doses.

In the 1980s, an approach was introduced that used the entire dose-response relationship for the chemical in establishing what were referred to as "benchmark doses." This approach has gained favor over the past several years, especially for chemicals with developmental or reproductive effects. This approach eliminates the need to identify a NOAEL, and generally reduces the uncertainty associated with developing a "safe" level. However, it requires more data than the NOAEL approach, and therefore more time and money.

In the benchmark dose (BMD) approach, the dose-response relationship is developed and a best-fit curve is identified based on the laboratory data (Figure 10.3). The data are not extrapolated to low-dose levels. Instead, the study is designed to bracket a certain frequency of developing toxicity. For example, let's assume we are conducting a reproductive study and the endpoint of interest is death of a fetus. We want to ensure that fetal death occurs no more than 10 percent of the time in laboratory rodents. This is referred to as an ED_{10} (i.e., effective dose in 10 percent of the population), and is shown in Figure 10.3. This is different from the NOAEL approach in that we are specifying the toxic endpoint and the level of protection that we wish to target. This provides for more relevant regulatory levels because they can change as conditions change.

Similar to cancer dose-response relationships (Figure 10.1), the best-fit curve steepness and location are uncertain. Therefore, a 95UCL curve is identified based on the variability of the laboratory data. An ED_{10} is also identified from this 95UCL curve. This lower (e.g., more health-protective) ED_{10} is then selected as the BMD. This is shown in Figure 10.3.

The highest NOAEL associated with this toxic effect might be either higher or lower than the BMD, depending on the potency of the chemical, the variability of the laboratory data, and many other factors. Depending on the levels, this approach might have little impact on the regulatory values currently used to protect human health and the environment. Nevertheless, this approach should be incorporated into the risk assessment process because it reflects best science and provides the most useful information upon which to base decisions.

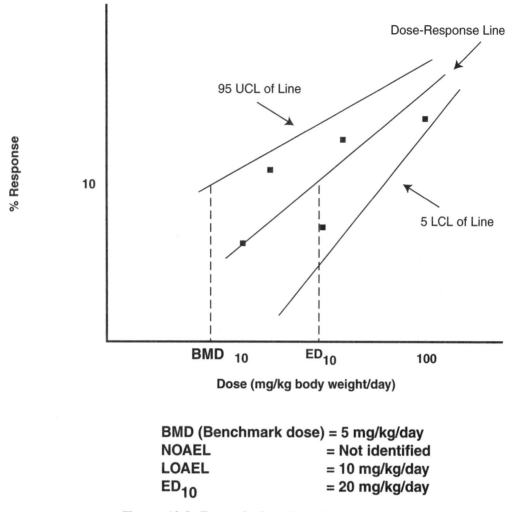

Figure 10.3. Example Benchmark Dose Duration

The BMD approach has been used for many years to identify acceptable surface water concentrations for exposure by fish and other aquatic species. The Ambient Water Quality Criteria (AWQC) adopted by the U.S. EPA are based on a series of laboratory tests that include identifying effective concentration levels for specific endpoints. These effective concentrations are then used to develop AWQC.

Probabilistic Approaches to Risk Assessment

The standardized approach for risk assessment discussed in Chapter 8 is called "deterministic" because the outcome of the risk assessment is a single risk, and only specific values are used to develop that risk. These specific values include assumptions that everyone has the same body weight, lives in one place for the same length of time, come into contact with the same amount of dirt, breathes the same amount of air, etc. In reality, individual variability exists that introduces uncertainty into the results obtained from using these simplifying assumptions.

One way to quantify this uncertainty is to develop a range of risks that corresponds with different likelihood of risk occurring across a population. This range of risks incorporates the known variability within the human population. Such an approach is called "probabilistic" because it is statistically based. This is a fairly complex mathematical process and typically involves hundreds of thousands of calculations by a computer that essentially mimics human variability. An overview of the process and its significance is presented here to familiarize you with the approach.

One probabilistic method that has become popular in recent years for use in risk assessments is called the Monte Carlo technique. Monte Carlo is the name of the mathematical formula used to generate the range of risks, and is not related to the location or gambling. This process is further discussed below.

Typically, risk assessments are conducted using exposure assumptions that represent upper-bound values of the known range. For example, an adult inhalation rate is assumed to be 20 m^3/day (0.8 m^3/hour). This value represents about the 95th percentile of all daily adult inhalation rates (Figure 10.4). In other words, 95 percent of all adults would breathe at a lower rate than this value. Because this estimate is high, it increases the amount of chemical exposure assumed by the inhalation route, which in turn leads to higher dose and risk. Increasing risk in this way actually diminishes the value of the result because it is known that the estimated risk is higher than the true risk. The question "how much higher?" can be answered through use of Monte Carlo methods.

To use computer models like Monte Carlo, the following values are needed:

- Body weight
- Soil ingestion rate
- Skin surface area
- Inhalation rate
- Exposure frequency
- Exposure duration
- Chemical concentration

Ranges for all of these parameters are somewhat defined for humans in the many surveys conducted by the government or private companies. The general distributions for several of these parameters are shown in Figure 10.4 and summarized in Table 10.2.

The Monte Carlo technique involves setting up dose and risk equations, and then having the computer solve the equation several thousand times (typically 10,000 iterations are run), each time randomly selecting from the distribution for each

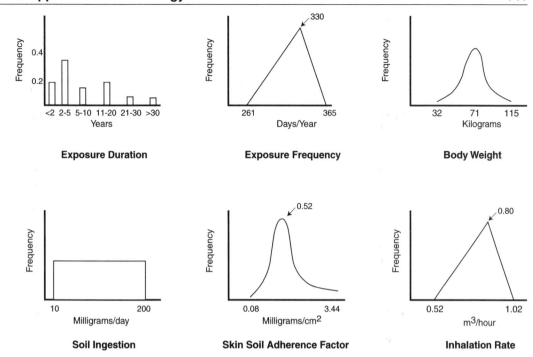

Figure 10.4. Sample Distributions of Exposure Assumptions in Human Populations

parameter. For example, to calculate the risk from inhaling chemical vapors, the body weight, inhalation rate, exposure duration and frequency, and concentration are needed. For each of these parameters, the computer selects from the distribution and uses the randomly selected value. This better reflects actual exposure because a range of intake assumptions is being used. The likelihood of obtaining any individual value is dependent on the shape of the distribution. For example, based on the shape of the distribution for body weights, the most frequent body weight used by the computer to solve the risk equation is likely around 155 pounds, which represents the most common value in the distribution. Only rarely would a body weight less than 80 pounds or greater than 300 pounds be seen. In the Monte Carlo simulation, these body weights would also only rarely occur.

The result of such a simulation is shown in Figure 10.5. In this graph, the calculated risk is plotted across the bottom and the frequency of that risk occurring is shown on the vertical axis. For this example, the average result is a cancer risk of 1×10^{-8}, which is very small (e.g., less than 1 in 100 million). Exactly half the time a lower risk would be expected, and half the time a higher risk would be expected. Another way of looking at this value is to say that such a value would only protect half the people because everyone to the right of this value on the graph has a higher risk. Therefore, a more protective level is usually targeted. This level is often the 95th upper percentile value, or sometimes the 99th percentile. Both of these levels are shown in Figure 10.5. At these percentile levels, only 5 percent and 1 percent of the population are likely to have a higher risk than that estimated. In this example, setting a

Table 10.2. Exposure Assumption Distributions

Exposure Assumption[a]	Distribution Assumed[b]	Relevant Values	Single Point Value[c]
Exposure duration	Custom	7 years (average), 30 years or more (top 11 percent)	30 years
Exposure frequency	Triangular	330 (most likely), 365 (maximum)	350 days/year
Exposure time	Custom	16 (50 percent), 24 (50 percent)	24 hours
Body weight	Normal	71 (average), 115 (maximum)	70 kg
Soil ingestion rate	Uniform	10 (minimum), 200 (maximum)	100 grams/day
Fraction of soil from contaminated area	Uniform	10 (minimum), 100 (maximum)	100 percent
Fraction of skin area exposed	Triangular	42 (most likely), 59 (maximum)	percent
Inhalation of dusts	Triangular	0.80 (most likely), 1.02 (maximum)	0.8 cubic meters/hour

[a] Parameters typically used to estimate exposure and risk for adult residents from chemicals in soil.
[b] Based on State of Ohio probabilistic risk assessment guidance.
[c] Used in typical deterministic risk assessments.

"safe" concentration at the 95th percentile leads to a risk level of 1×10^{-6} (one in one million), which is often targeted by regulatory agencies. In this case, probabilistic results from the risk assessment would indicate that estimated risks are below regulatory levels of concern. However, targeting the 99th percentile instead leads to a cancer risk estimate of 1×10^{-4} (one in 10,000), which will typically require cleanup to protect human health.

The single point estimate of risk using the typical deterministic approach can also be plotted on the graph. In this way the degree of conservatism built into the risk estimate from using upper-bound intake assumptions is quantified. In this example, the point estimate of risk is 1×10^{-5} (one in 100,000). This is three orders of magnitude higher risk than the average expected risk, and corresponds to about the 98th percentile. This indicates that the point estimate is protective of the population, but over-predicts risks for almost all possible combinations of exposure.

Such approaches can also be used for ecological risk assessment, which has higher levels of uncertainty in the risk estimates due to the lower amount of information on chemical exposure patterns by wildlife. Such approaches will likely not replace the deterministic methods of conducting risk assessment, but their utility lies in identifying how protective a concentration is likely to be under likely exposure conditions. Armed with this knowledge, regulatory agencies are better able to identify safe levels that protect the vast majority of the population. This increases the chances that cleanup costs will be used to reduce the potential risks from chemical exposure to acceptable levels, rather than cleaning a site up to levels much more restrictive than necessary to protect the public.

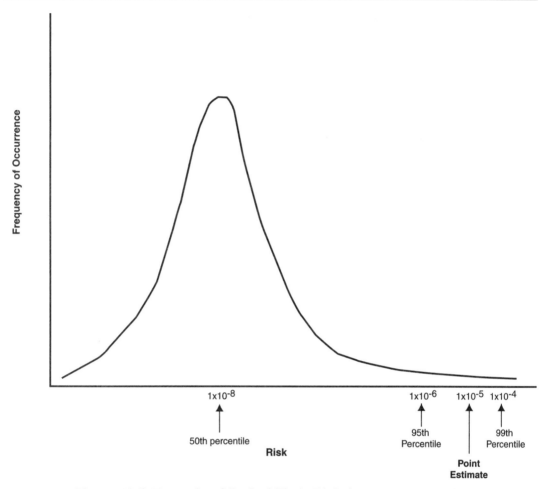

Figure 10.5. Example of Probabilistic Risk Assessment Results

Endocrine Disruptors and Beyond

Environmental estrogens are part of a group known as endocrine disruptors that can affect hormone action in an animal or person. This group of chemicals first gained attention in the 1980s when several environmental studies indicated that sex ratios of some reptiles and amphibians in specific areas were abnormal. Entire populations were composed primarily of females, and the males were feminized and not reproductively capable. DDT and dicofol (another organochloride pesticide) were known to be present in the water and sediment of their habitat, but at levels not considered to pose a health threat. It was later learned that the chemicals, passed from the mother into the egg, were impacting the developing embryo during the sexual differentiation process and preventing the embryo from developing into a male. This effect was very subtle and was not seen in the animals tested in the laboratory, but had dramatic impacts in the wild. Identifying a new type of toxicity that occurred at concentrations lower than those previously thought to have an effect was surprising and created alarm across communities because of its potential ramifications for wildlife and humans.

Today, research into endocrine disruptors is a very active branch of toxicology. It is relevant both for ecological receptors and humans, and may indicate that levels previously considered harmless are in fact causing harm. An overview of endocrine disruptors is provided below to illustrate how a new issue in toxicology can rapidly evolve into something that can impact our everyday lives.

A wide variety of chemicals have been shown to have endocrine disruption effects, including various types of pesticides, surfactants and detergents, metals, PCBs, dioxins, and PAHs (see Table 10.3). Recently phthalates, which are components of plastics, have been purported to cause endocrine disruption effects. The following types of effects have been linked to endocrine disruptors:

- Reproductive (feminization, demasculination, embryo death, failure to reproduce
- Developmental (deformities, hermaphroditism)
- Neurological (courtship behavior disruption, altered metabolic rate, abnormal sexual differentiation in the brain, teratogenicity of the central nervous system)
- Immunological (autoimmune syndromes, altered functions, reduced immune capacity)
- Thyroid disruption (hearing loss, hypothyroidism, decreased IQ)
- Cancer (liver, cervix, vagina, testes, breast, possible prostate)

The species and specific effects noted by chemical type are also shown in Table 10.3. Note that there are many other types of chemicals and causes of these effects. Many natural chemicals also have endocrine disruption effects, including chemicals in plants (phytoestrogens) and fungi (mycoestrogens). These natural endocrine disruptors are part of the natural system. Synthetic estrogens (e.g., manufactured chemicals that act like hormones) are additions to the system that can lead directly to toxic effects.

Hormones control many aspects of our lives, ranging from the fight or flight reaction of adrenaline to digesting sugars (e.g., insulin). Some hormones affect our reproductive systems, and it is these that were first identified in nature as endocrine disruptors. Because they affect so many of our processes, it is not surprising that their toxic effects can be very diverse.

One reason why endocrine disruptor effects are cause for concern is the extreme potency of hormones. Hormones are effective in our bodies at levels as low as 1 part per quadrillion (1 followed by24 zeroes). They are extremely short-lived; this is one reason they are potent. Our bodies are designed to respond to these chemicals at very low levels.

Table 10.3. Types of Chemicals Considered to Have Endocrine Disruptor Effects

Chemical Class	Example Chemicals	Affected Species
Polynuclear aromatic hydrocarbons (PAHs)	benzo(a)pyrene, anthracene	fish
Dioxins	dioxin (2,3,7,8-TCDD), dibenzofuran	monkeys, fish, mink, seals, birds, turtles
Polychorinated biphenyls (PCBs)	208 separate chemicals	dolphins, fish, mink, seals, humans, panthers, turtles
Metals	cadmium, lead, methylmercury, tributyltin	fish, panthers, snails, oysters
Organochloride insecticides	DDT, DDE, dicofol, lindane, toxaphene, aldrin	humans, rats, birds, fish, dolphins, alligators, panthers
Herbicides	simazine, molinate, trifluralin	rats, fish
Fungicides	benomyl, vinclozolin	fish, rats, mice, birds
Other insecticides	carbaryl, malathion, parathion, pentachlorophenol	birds, fish, rats, arthropods
Synthetic estrogens	diethylstilbestrol (DES)	humans
Surfactants (alkylphenol ethoxylates)	P-nonylphenol	fish

Environmental chemicals are not as potent as our own hormones, but they act by interfering in some way with our natural hormones. This interference can lead to differences in response to the hormone by the body. For example, a lake in Florida where DDT had been spilled several years before was found to contain no reproductively capable male crocodiles. The males had decreased levels of testosterone, had smaller gonads, and were unable to successfully breed. The females had higher levels of estrogens than at other locations. Crocodiles do not have sex chromosomes (e.g., x and y chromosomes). Instead, sexual differentiation is dependent on the temperature of the surrounding sediment. In the presence of other chemicals that can mimic hormonal action (e.g., estrogen), this sexual differentiation process can be altered or affected, leading to the types of males and females identified at this lake.

One chemical (beta-estradiol, or synthetic estrogen) has been shown to have endocrine disruption effects in humans. It is also by far the most potent endocrine disruptor identified. This should be expected because synthetic estrogen was designed to act like natural estrogen. However, many other chemicals not designed for such use also have similar effects, and also at low concentrations.

These types of sub-lethal effects that can occur in the environment at low levels remind us that we still know relatively little about the overall effects of chemicals introduced into our environment. As we have seen, there are over 50,000 chemicals in use today around the world. We have toxicological data on very few of these. It is unreasonable to assume that we have identified all of the toxic chemicals in use today, and the levels at which they are toxic to humans. Considering that chemicals can interact with each other in the environment and in the body—either increasing or decreasing the toxicity of each individual chemical—it is almost impossible to predict exactly what will happen from any specific chemical exposure. We find new chemicals every day, and many of these have specific and previously unknown toxicological mechanisms and effects.

Identifying a problem is often the most challenging part of solving it. Regarding endocrine disruptors, research is ongoing in an attempt to define the nature and extent of the problem (e.g., how many chemicals and how many species are or might be affected). Once the problem is defined, steps can be taken to reduce the problem by regulating use or release of the chemicals, or to eliminate the problem by developing new chemicals that do not have this serious side effect.

As much as we study toxicology, we still find new issues and problems we need to understand, such as with the endocrine disruptors discussed above. The more we learn, the more we find we have yet to learn. What new challenges will we face as we enter the 21st century? The Romans likely never knew the extent of the damage lead might have caused their society. Hopefully, through the science of toxicology and advances made over the past 50 years, we will not be similarly surprised.

Additional Reading

Kendall, R. J., R. L. Dickerson, J. P. Giesy, and W. P. Suk, 1998. *Principles and Processes for Evaluating Endocrine Disruption in Wildlife*. SETAC Technical Publications Series, SETAC Press.

Klaassen, C.D., 1996. *Casarett and Doull's Toxicology: The Basic Science of Poisons*. Fifth Edition. McGraw-Hill, New York.

Appendix A

Additional Reading

I. Regulatory Agency Publications

A. General Guidance

Ohio Environmental Protection Agency, 1996. Support Document for the Development of Generic Numerical Standards and Risk Assessment Procedures. The Voluntary Action Program, Division of Emergency and Remedial Response. (October 1996).

U.S. Environmental Protection Agency (U.S. EPA). 1989. Risk Assessment Guidance for Superfund, Volume I, Human Health Evaluation Manual (Part A), Interim Final. Office of Emergency and Remedial Response, Washington, D.C., EPA/540/1-89/002. (July 1989).

U.S. Environmental Protection Agency (U.S. EPA), 1986. Guidelines for Carcinogen Risk Assessment. *Federal Register* 51(185): 33992-34003.

U.S. Environmental Protection Agency (U.S. EPA), 1996. Proposed Guidelines for Carcinogen Risk Assessment; Notice. *Federal Register* 61(79): 17960-18011.

U.S. Environmental Protection Agency (U.S. EPA), 1997. Guiding Principles for Monte Carlo Analysis. Risk Assessment Forum, Washington, D.C. EPA/630/R-97/001. (March 1997).

U.S. Environmental Protection Agency (U.S. EPA), 1998. Guidelines for Ecological Risk Assessment. Risk Assessment Forum, Washington, D.C., EPA/630/R-95/002F. (April 1998).

B. Technical Information

California Environmental Protection Agency, 1998. Executive Summary for the Proposed Identification of Diesel Exhaust as a Toxic Air Contaminant. Air Resources Board, Office of Environmental Health Hazard Assessment. (April 22, 1998).

California Environmental Protection Agency, 1998. Findings of the Scientific Review Panel on The Report on Diesel Exhaust, as Adopted at the Panel's April 22, 1998 Meeting. Scientific Review Panel. (April 22, 1998).

International Agency for Research on Cancer (IARC), 1999. Overall Evaluations of Carcinogenicity to Humans. Located on the web at http://193.51.164.11/monoeval/crthall.html.

U.S. Environmental Protection Agency (U.S. EPA), 1993. Secondhand Smoke: What You Can do About Secondhand Smoke as Parents, Decisionmakers, and Building Occupants. Washington, D.C. EPA-402-F-93-004. (July 1993). Available on the web at http://www.epa.gov/iaq/pubs/tsbro.html.

II. Risk Assessment Publications

Anderson, E. L., and R. E. Albert, 1999. *Risk Assessment and Indoor Air Quality*. Lewis Publishers, Boca Raton, Florida. 272.

Daugherty, J. E., 1998. *Assessment of Chemical Exposures: Calculation Methods for Environmental Professionals*. Lewis Publishers, Boca Raton, Florida. 456.

Molak, V., 1997. *Fundamentals of Risk Analysis and Risk Management*. Lewis Publishers, Boca Raton, Florida. 496.

Neumann, D. A., and C. A. Kimmel, 1999. *Human Variability in Response to Chemical Exposures: Measures, Modeling, and Risk Assessment*. Lewis Publishers, Boca Raton, Florida. 272.

Sadar, A. J., and M. D. Shull, 1999. *Environmental Risk Communication: Principles and Practices for Industry*. Lewis Publishers, Boca Raton, Florida. 176.

Suter, G. W. II, L. W. Barnthouse, S. M. Bartell, T. Mill, D. MacKay, and S. Paterson, 1993. *Ecological Risk Assessment*. Lewis Publishers, Boca Raton, Florida.

III. Chemical- and Medium-specific Toxicology

Greek, K., 1981. *Carciac Glycosides*. Springer-Verlag, New York.

Halstead, B. W., 1965. *Poisonous and Venomous Marine Animals*, Volumes 1 and 2. U.S. Government Printing Office, Washington, D.C.

Hayes, W. J. Jr., 1982. *Pesticide Studies in Man*. Williams and Wilkins Company, Baltimore, Maryland.

Gilfillin, S. C., 1965. Lead Poisoning and the Fall of Rome. *Journal of Occupational Medicine* 7: 53-60.

Kendall, R. J., R. L. Dickerson, J. P. Giesy, and W. P. Suk, 1998. *Principles and Processes for Evaluating Endocrine Disruption in Wildlife*. SETAC Technical Publications Series, SETAC Press.

Lee, D. K. H., 1972. *Metallic Contaminants and Human Health*. Academic Press, New York. 241.

Nriago, J. O., 1983. *Lead and Lead Poisoning in Antiquity*. John Wiley and Sons, New York. 437.

Olin, S. S., 1999. *Exposure to Contaminants in Drinking Water*. Lewis Publishers, Boca Raton, Florida. 256.

Tucker, A., 1972. *The Toxic Metals*. Ballantine Books, Inc., New York. 237.

Waldron, H. A., 1973. Lead Poisoning in the Ancient World. *Medical History* 17: 391-399.

Weinberg, 1983. A Molecular Basis of Cancer. *Scientific American*. December, 1983.

IV. General Toxicology Technical Publications

Conway, R. A., 1982. *Environmental Risk Analysis for Chemicals*. Van Nostrand Reinhold, New York.

Hall, S. K., J. Chakraborty, and R. J. Ruch, 1997. *Chemical Exposure and Toxic Responses*. Lewis Publishers, Boca Raton, Florida. 304.

Hayes, A.W., 1984. *Principles and Methods of Toxicology*. Raven Press, New York. 750.

Hodgson, E., and F. E. Guthrie, 1984. *Introduction to Biochemical Toxicology*. Elsevier Science Publishing Co., Inc., New York, NY. 437.

Klaassen, C. D., 1996. *Casarett and Doull's Toxicology: The Basic Science of Poisons*. Fifth Edition. McGraw-Hill, New York.

Philip, R. B., 1995. *Environmental Hazards and Human Health*. Lewis Publishers, Boca Raton, Florida. 320.

V. Environmental Chemistry and Toxicology

American Chemical Society, 1978. *Cleaning Our Environment: A Chemical Perspective*. Second edition. ACS, Washington, D.C.

Bailey, R. A., H. M. Clarke, J. P. Ferris, S. Krause, and R. L. Strong, 1978. *Chemistry of the Environment*. Academic Press, New York.

Godish, T., 1998. *Air Quality*. Third Edition. Lewis Publishers, Boca Raton, Florida. 464.

Stern, A. C., et al., 1973. *Fundamentals of Air Pollution*. Academic Press, New York.

Thibodeaux, L. J., 1979. *Chemodynamics: Environmental Movement of Chemicals in Air, Water, and Soil*. Wiley-Interscience, New York.

VI. General Interest

Carson, R., 1962. *Silent Spring*. Houghton Mifflin Company, Boston, Massachusetts.

Harr, J., 1995. *A Civil Action*. Vintage Books, A Division of Random House, Inc., New York. 502.

Glossary

Absorption	Passage of a chemical across a membrane and into the body.
Acaricide	A pesticide that targets spiders.
Acceptable Risk	A risk that is so low that no significant potential for toxicity exists, or a risk society considers is outweighed by benefits.
Acute	A single or short-term exposure period.
Alkaloid	A diverse group of structurally related chemicals naturally produced by plants; many of these chemicals have high toxicity.
Ames Assay	A popular laboratory *in vitro* test for mutagenicity using bacteria.
Anesthetic	A toxic depressant effect on the central nervous system.
Assessment Endpoint	An ecological value representing the focus for protection in an ecological risk assessment.
Background	Frequency at which toxic effects occur at naturally occurring levels (e.g., cancer).
Benchmark Dose	A dose that is intended to protect a specific percentage of the population from a specific toxic endpoint. This typically represents the 95UCL of an effective dose to 10% of a test population (ED_{10}); this dose would therefore protect 90 percent of the population.

Bioassay	A toxicity study in which specific toxic effects from chemical exposure are measured in the laboratory using living organisms.
Cancer	Uncontrolled cell division, exhibited by tumor formation.
Carcinogen	A cancer-causing substance.
Catalyst	A chemical that lowers the energy needed for an enzymatic reaction to occur.
Chloracne	A toxic effect of hydrocarbons containing halogen atoms (e.g., chlorine) characterized by severe acne around the ears, eyes, shoulders, back, and genitalia.
Chronic	An exposure period encompassing the majority of the life span for a laboratory animal species, or covering at least 10 percent of a human's life span.
Cofactor	A chemical that is required by enzymes for certain metabolic reactions to occur.
Community	An interacting group of organisms residing in a common environment.
Conceptual Site Model	A description and schematic diagram illustrating the relationships between chemicals in the environment from a specific source and potential receptors.
Cytochrome P450	A group of structurally related enzymes comprising the mixed function oxidase (MFO) system; responsible for initial metabolism of a wide variety of foreign chemicals.
Delaney Clause	A 1958 addition to the Federal Food, Drug, and Cosmetic Act which bans any substance from use in food that is shown to cause cancer in mice.
Dermal Contact	Exposure to a chemical through the skin.
Deterministic Assessment	A risk assessment using single-point estimates of exposure parameters.

Developmental Toxicity	Toxic, sublethal effects on a developing fetus.
Disposition	The toxicokinetics of a chemical within the body.
Distribution	A typically quantitative description of where a chemical goes in the body following absorption.
DNA	Deoxyribonucleic Acid.
Dose	The amount of a chemical entering the body per unit time period.
Dose-Response	The relationship between the amount of a chemical taken into the body and the degree of toxic response.
Draize Test	An acute toxicity test designed to test chemicals for potential eye irritation; usually conducted in rabbits.
Ecosystem	The sum total of all interactions linking organisms in a community with each other and their environment.
ED_{10}	Dose level at which 10 percent of exposed individuals have a specific toxic response.
Efficacy	The relative maximum effect that can result from any dose level of a chemical.
Endocrine Disruptor	A chemical that affects the normal functioning of the endocrine system.
Environmental Toxicology	The branch of toxicology focusing on the impact of chemical pollutants in the environment on biological organisms.
Enzyme	A protein that increases the rate of a metabolic reaction.
Epidemiology	The study of the incidence, prevalence, source, and cause of disease in large populations.
Epigenetic	A substance that causes cancer without directly acting on DNA.
Excretion	Elimination of a chemical from the body.

Exposure	Contact with a chemical by a living organism.
Exposure Assessment	The component of a human health risk assessment in which a conceptual site model and chemical-specific dose estimates are made.
Exposure Characterization	The component of an ecological risk assessment in which exposure is evaluated. This is parallel to the exposure assessment component of a human health risk assessment.
Exposure Duration	The number of years a receptor is exposed to a chemical. This can be either a measured or assumed value.
Exposure Frequency	The number of days per year a receptor is exposed to a chemical. This can be either a measured or assumed value.
Exposure Medium	A matrix (e.g., soil, water, air) in which an organism comes into contact with a chemical.
Exposure Pathway	The environmental course a chemical takes from a source to an organism.
Exposure Point	A location where a receptor is exposed to a chemical. This can be either a known or assumed location.
Exposure Profile	The product of the exposure characterization component of an ecological risk assessment in which exposure is estimated or measured.
Exposure Route	The way in which a chemical comes in contact with an organism (e.g., ingestion).
Extrapolation	Using data from direct observations, typically laboratory animal tests, to predict results for unobserved conditions.
FDA	United States Food and Drug Administration.
Fetotoxic	A chemical that causes toxic effects in a fetus in utero.
FIFRA	United States Federal Insecticide, Fungicide, and Rodenticide Act.

First-Pass Effect	Chemical transformation in the liver following absorption from the gastrointestinal tract prior to release of the chemical to other locations.
Food Additive	A manufactured chemical added to a food product.
FD&C Act	United States Food, Drug, and Cosmetic Act.
Food Toxicology	The branch of toxicology focusing on the effects of chemical substances in food.
Fumigant	A chemical used as a vapor to control insects, nematodes, weeds, and fungi in soil, and consumables stored in confined spaces (e.g., grain in silos).
Fungicide	A pesticide targeted on fungi and molds.
GRAS	A term used by the FDA for food additives that are "generally recognized as safe."
Hazard	Degree of likelihood of non-cancer adverse effects occurring from chemical exposure.
Hazard Quotient	The ratio of a measured or estimated dose to a chemical-specific reference dose through a specific route of exposure.
Hazard Index	The sum of hazard quotients across multiple chemicals and/or multiple exposure routes.
Herbicide	A pesticide that targets plants.
Hormone	A chemical produced by cells in one part of an organism that affect the functioning of other locations within the organism.
IARC	The International Agency for Research on Cancer.
Ingestion	Exposure to a chemical through the mouth.
Inhalation	Exposure to a chemical through the respiratory system.
Initiation	The mutation of a gene; the first step in the process of carcinogenesis.

Insecticide	A pesticide that targets insects.
Intake Assumption	A value used to represent one parameter in the quantitative estimation of exposure. This can either be measured or estimated.
in vitro	Latin term meaning "in glass." Used to describe toxicity tests using only cell or tissue cultures.
Involuntary Risk	A risk that is not under personal control.
Irreversible Effect	A toxic effect that is permanent, even if exposure to the chemical causing the effect ceases.
Kilogram	One thousand grams.
Latency Period	The time interval between initial exposure to a cancer-causing agent and development of the cancer.
LC_{50}	Concentration of a chemical lethal to 50 percent of organisms in a laboratory study.
LD_{50}	Dose of a chemical lethal to 50 percent of organisms in a laboratory study.
Leaching	Movement of a chemical from soil or sediment to water (surface or groundwater).
Leukemia	Cancer of the bone marrow; primarily affects white blood cells.
Linearized Multistage Model	Most common model used by U.S. EPA to develop slope factors; assumes a linear dose-response relationship and no threshold for effects.
Lipid	Fat tissue.
Lipophilic	Fat-loving; the degree to which a chemical partitions into fat versus water.
LOAEL	Lowest-observed adverse effect level; the lowest dose at which an adverse effect is seen.
LOEL	Lowest-observed effect level (not necessarily a toxic effect).
Lymphoma	A tumor originating in lymph nodes.

Maximum Tolerated Dose (MTD)	The highest dose that does not cause death in a laboratory study.
Measure of Effect	Quantitative estimates or measurements of toxicity used in an ecological risk assessment to properly evaluate management goals and assessment endpoints.
Melanoma	A type of skin tumor that affects melanocytes.
Mesothelioma	A form of lung cancer that, so far, is unique to asbestos exposure in humans.
Metabolic Activation	Formation of a more toxic chemical through metabolism of the parent chemical.
Metabolic Saturation	Limiting dose of a chemical that can be detoxified through metabolic reactions involving enzymes and cofactors.
Metabolism	Transformation of a chemical within an organism to other chemicals through reactions.
Micrograms	One millionth of a gram (0.000001 gram).
Milligrams	One thousandth of a gram (0.001 gram).
Mixed Function Oxidase (MFO)	An important group of Phase I enzymes that metabolize a wide variety of chemicals (see Cytochrome P450).
Molluscicide	A pesticide that targets molluscs.
Mutagen	A chemical that causes a change in DNA.
Mutation	A change in normal DNA structure.
Mycotoxin	A toxic chemical produced by a fungus or mold.
Nematocide	A pesticide that targets roundworms (Nematodes).
Neurological	Pertaining to the nervous system.
Neurotoxin	A chemical that causes an effect on the nervous system.
NOAEL	No-observed adverse effect level; the highest dose at which no adverse effects are seen.

NOEL	No-observed effect level; the highest dose at which no effects are seen.
Organic	A chemical containing carbon.
Organophosphorus Insecticide	An insecticide that exerts its toxic effect by interfering with acetylcholinesterase (an enzyme that breaks down the neurotransmitter acetylcholine).
PAHs	Polynuclear aromatic hydrocarbons; naturally occurring chemicals formed as a result of incomplete combustion. Several of the chemicals in this group are carcinogenic.
PCBs	Polychlorinated biphenyls; a manufactured group of chemicals used for their insulating properties. Several of the chemicals in this group are carcinogenic.
Pesticide	A chemical used to control pests.
Pollution	Introduction of chemicals into the environment from human activities.
Population	A group of organisms within a species that interbreed.
Potency	The relative degree of toxic effects caused by a chemical at a specific dose.
Probabilistic Assessment	A risk assessment involving the use of distributions of exposure parameters, and resulting in a distribution of possible risks; incorporates variability into a risk assessment.
Problem Formulation	The initial stage of an ecological risk assessment where the problem is defined and the scope and endpoints of the assessment are identified.
Progression	The series of events that occur over time whereby cancer cells form a tumor; the final step in the process of carcinogenesis.
Promotion	The process whereby mutant cells become cancer cells through abnormal cell division; the third step in the process of carcinogenesis.

Receptor	A risk assessment term describing a hypothetical individual or group of organisms assumed to be exposed to a chemical.
Redistribution	The transfer of a chemical from one tissue or organ to another over time (e.g., liver to bone).
Reference Dose	A toxicity value used in human health risk assessments representing a "safe" dose (i.e., not associated with any toxic effect).
Release Mechanism	The manner in which a chemical moves from a source location to an exposure medium.
Reproductive Toxicity	Toxic effects impacting the ability of an organism to reproduce.
Reversible Effect	An effect that dissipates with time following cessation of chemical exposure.
Risk	The probability of an adverse effect resulting from an activity or from chemical exposure under specific conditions.
Risk Assessment	A scientific process used to estimate possible exposures, cancer risks, and non-cancer adverse health effects from known or estimated levels of chemicals.
Risk Characterization	The component of a human health or ecological risk assessment where cancer risks and non-cancer adverse health effects are calculated using mathematical equations.
Risk Communication	The process of making risk assessment and risk management information understandable to the layperson.
Risk Description	A portion of the risk characterization component of an ecological risk assessment in which the significance of any estimated adverse effect is interpreted from the lines of evidence used to estimate risk.
Risk Estimation	A portion of the risk characterization component of an ecological risk assessment in which exposure and effects data are compared. This

is parallel to the risk characterization component of a human health risk assessment.

Risk Management	The manner in which identified and/or estimated chemical risks are controlled (e.g., lowered) at a given location.
Rodenticide	A pesticide that targets rodents.
Sensitivity	The intrinsic degree of an individual's susceptibility to a specific toxic effect.
Sensitization	The process of becoming more sensitive to a toxic response from initial exposure to the causative agent.
Slope Factor	A plausible, upper bound estimate of the probability of an individual developing cancer over a lifetime of chemical exposure.
Subchronic Exposure	An exposure period intermediate between acute and chronic; typically 90 days in laboratory rodent studies.
Target Organ	The primary organ where a chemical causes non-cancer toxic effects.
Teratogen	A chemical causing a mutation in the DNA of a developing offspring.
Threshold Dose	A dose below which no adverse effects will occur.
Toxin	A biologically-produced chemical that has toxicity.
Toxic Effect	An adverse impact caused by a chemical.
Toxicity	The intrinsic degree to which a chemical causes adverse effects.
Toxicity Assessment	The component of a human health risk assessment where slope factors and reference doses are developed and identified.
Toxicodynamics	How a chemical interacts with its target site; typically focuses on chemical interactions and mechanism of action.

Toxicokinetics	The manner in which a chemical acts in an organism; typically focuses on the concentration of a chemical in tissues over time.
Toxicology	The study of poisons and their action on living systems.
Transport Medium	A medium (i.e., water, air, or soil) through which chemicals can move to locations remote from their source.
TSCA	United States Toxic Substances Control Act.
Tumor	An uncontrolled, abnormal growth of cells in any tissue.
Unacceptable Risk	A risk that is perceived to be high enough to represent a significant potential for toxicity, or a risk society considers is not outweighed by benefits.
Uncertainty	Error introduced into an evaluation due to lack of knowledge.
Uncertainty Factor	Values used to account for lack of knowledge when conducting extrapolation.
95UCL	The upper 95th percentile of the arithmetic mean; the true mean is only likely to be greater than this value 5% of the time.
U.S. EPA	United States Environmental Protection Agency.
Variability	Error introduced into an evaluation due to heterogeneity of populations.
Volatilization	Upward movement of chemicals in the vapor phase from soil or water into air.
Voluntary Risk	A risk considered acceptable because it is chosen.
WHO	World Health Organization.
Xenobiotic	A chemical foreign to a living organism.

Index

Government Institutes Mini-Catalog

PC #	ENVIRONMENTAL TITLES	Pub Date	Price
629	ABCs of Environmental Regulation: Understanding the Fed Regs	1998	$49
627	ABCs of Environmental Science	1998	$39
672	Book of Lists for Regulated Hazardous Substances, 9th Edition	1999	$79
579	Brownfields Redevelopment	1998	$79
4100	CFR Chemical Lists on CD ROM, 1998 Edition	1997	$125
4089	Chemical Data for Workplace Sampling & Analysis, Single User Disk	1997	$125
512	Clean Water Handbook, 2nd Edition	1996	$89
581	EH&S Auditing Made Easy	1997	$79
673	E H & S CFR Training Requirements, 4th Edition	1999	$89
4082	EMMI-Envl Monitoring Methods Index for Windows-Network	1997	$537
4082	EMMI-Envl Monitoring Methods Index for Windows-Single User	1997	$179
525	Environmental Audits, 7th Edition	1996	$79
548	Environmental Engineering and Science: An Introduction	1997	$79
643	Environmental Guide to the Internet, 4rd Edition	1998	$59
650	Environmental Law Handbook, 15th Edition	1999	$89
353	Environmental Regulatory Glossary, 6th Edition	1993	$79
652	Environmental Statutes, 1999 Edition	1999	$79
4097	OSHA CFRs Made Easy (29 CFRs)/CD ROM	1998	$129
4102	1999 Title 21 Food & Drug CFRs on CD ROM-Single User	1999	$325
4099	Environmental Statutes on CD ROM for Windows-Single User	1997	$139
570	Environmentalism at the Crossroads	1995	$39
536	ESAs Made Easy	1996	$59
515	Industrial Environmental Management: A Practical Approach	1996	$79
510	ISO 14000: Understanding Environmental Standards	1996	$69
551	ISO 14001: An Executive Report	1996	$55
588	International Environmental Auditing	1998	$149
518	Lead Regulation Handbook	1996	$79
554	Property Rights: Understanding Government Takings	1997	$79
582	Recycling & Waste Mgmt Guide to the Internet	1997	$49
615	Risk Management Planning Handbook	1998	$89
603	Superfund Manual, 6th Edition	1997	$115
566	TSCA Handbook, 3rd Edition	1997	$95
534	Wetland Mitigation: Mitigation Banking and Other Strategies	1997	$75

PC #	SAFETY and HEALTH TITLES	Pub Date	Price
547	Construction Safety Handbook	1996	$79
553	Cumulative Trauma Disorders	1997	$59
663	Forklift Safety, 2nd Edition	1999	$69
539	Fundamentals of Occupational Safety & Health	1996	$49
612	HAZWOPER Incident Command	1998	$59
535	Making Sense of OSHA Compliance	1997	$59
589	Managing Fatigue in Transportation, ATA Conference	1997	$75
558	PPE Made Easy	1998	$79
598	Project Mgmt for E H & S Professionals	1997	$59
552	Safety & Health in Agriculture, Forestry and Fisheries	1997	$125
669	Safety & Health on the Internet, 3rd Edition	1999	$59
597	Safety Is A People Business	1997	$49
668	Safety Made Easy, 2nd Edition	1999	$59
590	Your Company Safety and Health Manual	1997	$79

Government Institutes

4 Research Place, Suite 200 • Rockville, MD 20850-3226
Tel. (301) 921-2323 • FAX (301) 921-0264
Email: giinfo@govinst.com • Internet: http://www.govinst.com

Please call our customer service department at (301) 921-2323 for a free publications catalog.

CFRs now available online. Call (301) 921-2355 for info.

Government Institutes Order Form

4 Research Place, Suite 200 • Rockville, MD 20850-3226
Tel (301) 921-2323 • Fax (301) 921-0264
Internet: http://www.govinst.com • E-mail: giinfo@govinst.com

4 EASY WAYS TO ORDER

1. Tel: **(301) 921-2323**
Have your credit card ready when you call.

2. Fax: **(301) 921-0264**
Fax this completed order form with your company purchase order or credit card information.

3. Mail: **Government Institutes Division**
ABS Group Inc.
P.O. Box 846304
Dallas, TX 75284-6304   USA
Mail this completed order form with a check, company purchase order, or credit card information.

4. Online: Visit http://www.govinst.com

PAYMENT OPTIONS

❑ **Check** (payable in US dollars to **ABS Group Inc.** **Government Institutes Division**)

❑ **Purchase Order** (This order form must be attached to your company P.O. Note: All International orders must be prepaid.)

❑ **Credit Card** ❑ VISA ❑ MasterCard ❑ AMERICAN EXPRESS

Exp. ____ /____

Credit Card No. _____

Signature _____

(Government Institutes' Federal I.D.# is 13-2695912)

CUSTOMER INFORMATION

Ship To: (Please attach your purchase order)

Name _____

GI Account # (7 digits on mailing label) _____

Company/Institution _____

Address _____
(Please supply street address for UPS shipping)
_____

City _____ State/Province _____

Zip/Postal Code _____ Country _____

Tel ( ) _____

Fax ( ) _____

Email Address _____

Bill To: (if different from ship-to address)

Name _____

Title/Position _____

Company/Institution _____

Address _____
(Please supply street address for UPS shipping)
_____

City _____ State/Province _____

Zip/Postal Code _____ Country _____

Tel ( ) _____

Fax ( ) _____

Email Address _____

Qty.	Product Code	Title	Price

❑ **New Edition No Obligation Standing Order Program**
Please enroll me in this program for the products I have ordered. Government Institutes will notify me of new editions by sending me an invoice. I understand that there is no obligation to purchase the product. This invoice is simply my reminder that a new edition has been released.

15 DAY MONEY-BACK GUARANTEE
If you're not completely satisfied with any product, return it undamaged within 15 days for a full and immediate refund on the price of the product.

SOURCE CODE: BP01

Subtotal _____
See below for appropriate Sales Tax _____
See below for Shipping and Handling _____
Total Payment Enclosed _____

Shipping and Handling	Sales Tax
Within U.S: 1-4 products: $6/product; 5 or more: $4/product	Maryland 5%
	Tennessee 6%
Outside U.S: Add $15 for each item (Global)	Texas 8.25%
	Virginia 4.5%